Environmental Footprints and Eco-design of Products and Processes

Series Editor

Subramanian Senthilkannan Muthu, Head of Sustainability - SgT Group and API, Hong Kong, Kowloon, Hong Kong

Indexed by Scopus

This series aims to broadly cover all the aspects related to environmental assessment of products, development of environmental and ecological indicators and eco-design of various products and processes. Below are the areas fall under the aims and scope of this series, but not limited to: Environmental Life Cycle Assessment; Social Life Cycle Assessment; Organizational and Product Carbon Footprints; Ecological, Energy and Water Footprints; Life cycle costing; Environmental and sustainable indicators; Environmental impact assessment methods and tools; Eco-design (sustainable design) aspects and tools; Biodegradation studies; Recycling; Solid waste management; Environmental and social audits; Green Purchasing and tools; Product environmental footprints; Environmental management standards and regulations; Eco-labels; Green Claims and green washing; Assessment of sustainability aspects.

More information about this series at http://www.springer.com/series/13340

Subramanian Senthilkannan Muthu
Editor

Circular Economy

Assessment and Case Studies

Editor
Subramanian Senthilkannan Muthu
Head of Sustainability
SgT and API
Kowloon, Hong Kong

ISSN 2345-7651 ISSN 2345-766X (electronic)
Environmental Footprints and Eco-design of Products and Processes
ISBN 978-981-16-3700-1 ISBN 978-981-16-3698-1 (eBook)
https://doi.org/10.1007/978-981-16-3698-1

This Springer imprint is published by the registered company Springer Nature Singapore Pte Ltd.
The registered company address is: 152 Beach Road, #21-01/04 Gateway East, Singapore 189721, Singapore

Contents

About the Editor

Dr. Subramanian Senthilkannan Muthu currently works for SgT Group as Head of Sustainability, based in Hong Kong. He earned his PhD from The Hong Kong Polytechnic University and is a renowned expert in the areas of environmental sustainability in textiles and clothing supply chain, product life cycle assessment (LCA), ecological footprint and product carbon footprint assessment (PCF) in various industrial sectors. He has five years of industrial experience in textile manufacturing, research and development and textile testing and over a decades of experience in life cycle assessment (LCA), and carbon and ecological footprint assessment of various consumer products. He has published more than 100 research publications, written numerous chapters and authored/edited over 100 books in the areas of carbon footprint, recycling, environmental assessment and environmental sustainability.

Brp—Exploring Paths to the Circular Economy

Sabrina Roy-Pelletier and Emmanuel Raufflet

Abstract Sherbrooke, Canada, October 2018—A team of managers of BRP (Bombardier Recreational Products) attended a presentation on circular economy. The following day, they decided to explore the potential of the circular economy model for their product lines **(see Appendix 1: Definition of circular economy)**. BRP defined itself as a company *that creates innovative ways to move—on snow, water, on asphalt or dirt—even in the air* (https://www.brp.com/en/about-brp.html, April 29 2019). Its portfolio of products included snowmobiles, watercraft, off-road vehicles, boats, pontoons, marine propulsion systems (motors), engines for karts, motorcycles and recreational aircrafts. Several of these outdoors, motor-driven vehicles had been criticized for their environmental and noise impacts. BRP had set sustainable development goals to address these challenges. RP management had set the goal to become the leader in sustainability practices and innovation in the powersports industry by 2020 ("Corporate Social Responibility—Fiscal Year 2018" https://www.brp.com/content/dam/corpo/Global/Documents/csr documents/BRP_Overview%20GRI2018_EN_final.pdf, November 19, 2018). A lot had been achieved in sustainability per se in the last decade. However, evaluating these sustainable development achievements with the lens of circular economy needed to be done. The team had drafted several questions for this exploratory exercise. Among them: What relevant learning could be generated from the application of the lenses of circular strategies to business models, value chain, and operations? Based on this analysis which circular strategies should be prioritized? The team started this initial exploratory exercise aware that more information was needed to define implementation priorities.

Chapter/Case submitted to Muthu, S. S. et al. (forthcoming), Circular Economy-Assessment, Case Studies- Springer-Nature Publications

S. Roy-Pelletier (✉) · E. Raufflet
HEC Montréal, Montreal, Canada
e-mail: sabrina.roy-pelletier@hec.ca

S. S. Muthu (ed.), *Circular Economy*, Environmental Footprints and Eco-design of Products and Processes, https://doi.org/10.1007/978-981-16-3698-1_1

1 History

In 1922, Joseph-Armand Bombardier, a young mechanic from Valcourt, Quebec, designed the first tracked vehicle to travel on snow. In 1937, he patented the B7 snowmobile, his first major commercial success. The product was designed mainly for the very snowy streets of Quebec, which were not cleared at the time. The intention of this entrepreneur was to make it possible to travel in winter and reduce the isolation of Canadian rural communities.[1]

Over decades as customer needed change, the company innovated to adapt its products. In 1949, as the widespread use of snow removal in winter reduced demand for snowmobiles, Bombardier developed vehicles for the forestry industry based on an interchangeable system of wheels and skids. In 1959, the company launched its flagship product, the Ski-Doo, a light track recreational snowmobile for two passengers. Sales of the Ski-Doo jumped from 225 units in 1959 to 8,210 in four years and it quickly became a major source of sales for Bombardier, with revenues rising from $10 million in 1964 to $183 million in 1972. In 1970, Bombardier acquired Austria-based Lohnerwerke, specialized in the manufacturing of trams. Lohnerwerke had a subsidiary named Rotax, which manufactured engines for Ski-Doo.[2]

Bombardier further diversified its activities in 1974 as it was awarded the contract to manufacture 423 cars for the Montreal metro through Lohnerwerke-based competencies. In 1986, Bombardier acquired the state-owned aeronautics company Canadair. This marked the beginning of growth through acquisitions in the rail and aerospace industries to integrate their expertise within Bombardier.

At the same time, the company was developing new recreational products such as the first personal watercraft in 1968 under the Sea-Doo brand, a jet-propelled boat in 1994, an all-terrain vehicle (ATV) prototype in 1998 and many other models and variations of its various recreational vehicles.

In 1996, Bombardier's product structure was divided into three main divisions: aeronautics, rail transportation and leisure products. In 2003, Bombardier faced a difficult situation due to the crisis in the aviation industry following September 11, 2001. President and CEO Paul Tellier sold the Recreational Vehicle Division (RVB) to generate cash. The BRP brand was then officially created in 2004 as a company independent from Bombardier.[3]

An independent entity, BRP continued to develop various products, such as the Spyder on-road vehicle in 2007, the side-by-side off-road vehicle (SSV) in 2010, as well as several other innovations and technological advances related to motorsports.

---

[1] Jonathan McQuarrie, "Bombardier Inc.", www.thecanadianencyclopedia.ca, November 17, 2018.

[2] Jonathan McQuarrie, *op. cit.*

[3] "Plus de 75 ans d'innovation" https://www.brp.com/fr/a-propos-de-brp/our-story.html, November 17, 2018.

2 BRP: Products

As of 2018, BRP was still known for its iconic product, the snowmobile with 101 different models. It was also a world leader and one of the most diversified manufacturers of motorized sports vehicles, offering products such as personal watercraft, off-road vehicles, on-road vehicles, boats, and engines and accessories in more than 100 countries.[4] Products are divided into four main sectors: propulsion systems (Evinrude and Rotax engines), PAC (parts, accessories and clothing), all-season products (ATVs, SSVs and Spyders) and seasonal products (Ski-Doo and Lynx snowmobiles and Sea-Doo personal watercraft).[5] In 2018, BRP's largest market was the United States (50.5% of revenues), followed by Canada (17.3% of revenues).

BRP production was carried out by eight plants[6]:

- One in Canada near the head office in Valcourt, which manufactured Ski-Doo snowmobiles, Spyder vehicles and Rotax engines,
- Two in the United States for the production and assembly of Evinrude outboard engines and Rotax propulsion systems,
- Three in Mexico for the manufacture of ATV and SSV vehicles, the assembly of Sea-Doo personal watercraft and Rotax engines,
- One in Finland for the assembly of Lynx and Ski-Doo snowmobiles as well as some specialized ATVs,
- One in Austria for Rotax engines.

3 Competition

Five manufacturers, namely BRP, Textron Off Road/Arctic Cat, Polaris, Honda and Yamaha dominated the recreational motor vehicle market:

As for the SSV market, BRP's main competitor was Polaris, with almost 40% of the market share, compared to 13% for BRP.[7]

As for the ATV market, the main competitors were Polaris and Honda, which hold 33 and 29% of the market share respectively, compared to 15% for BRP.[8] Global market trends between 2012 and 2017 indicated that sales of snowmobiles, ATVs and Spyder-type vehicles were declining over the entire period, compared to personal watercraft and ISS. In the United States, market were forecasted a yearly decrease

---

[4] BRP Inc., (2018) "Annual information form: fiscal year ended January 31, 2018".

[5] Ibid.

[6] BRP Inc., (2018) "Annual information form: fiscal year ended January 31, 2018".

[7] "Projected Global Side-by-side All-terrain Vehicle Market Share in 2018, by Vendor." April 3, 2019, www.statista.com/statistics/438113/global-side-by-side-atv-market-share/.

[8] "Share of Americans Who Own An Atv (All-terrain Vehicle) in 2018, by Age." April 12, 2019, www.statista.com/statistics/228874/people-living-in-households-that-own-an-atv-all-terrain-vehicle-usa/.

of 0.2% between 2018 and 2023 in off-road vehicle sales.[9] Regarding the Canadian recreational vehicle market, a slowdown in industry growth was expected by 2023, due to the negative impact of a 4.4% increase in gasoline prices and a 1.1% increase in electricity prices.[10]

4 Key Success Factors[11]

Key success factors in this industry included:

- Capacity to design and manufacture products compliant with legal standards;
- Mass production to take advantage of economies of scale to manufacture more efficiently and increase profit margins;
- Access to the most recent and efficient technologies and techniques to offer a quality and competitive product;
- Product offer popular with the market, including diverse client preferences and weather conditions;
- The ability to adapt to changes in customer preferences.

5 BRP's Strategic Priorities

BRP had identified three strategic priorities for 2018[12]:

- The economic growth of the company;
- The "agile" transformation of the supply chain to improve the customer experience and;
- Reducing working capital and improving business processes to make the company leaner.

6 Brp Value Chain

BRP's simplified value chain is as follows: BRP develops new vehicle models, procures raw materials and components in order to be able to manufacture them, produces the vehicles and sells them to dealers. Dealers distribute them to customers. Customers use them until the end of the vehicles' life.

---

[9] Ediz Ozelkan (2018), "ATV, Golf Cart & Snowmobile Manufacturing in Canada", IBISWorld Industry Report.

[10] Devin Savaskan, (2018), "ATV Manufacturing in the US ", IBISWorld Industry Report.

[11] Ediz Ozelkan (2018), op. cit.

[12] BRP Inc., (2018) "Annual information form: fiscal year ended January 31, 2018".

7 Use of Raw Materials and Components

In the snowmobile and off-road vehicle industry, purchases of raw materials and components accounted for 59.2% of total costs in 2018, compared to 52.2% in 2013. The average is 54.2% for all industries combined in 2018. Purchase prices in the recreational vehicle industry are highly volatile, as they depend on market demand, as is the price of aluminum, which rose sharply between 2013 and 2018.[13]

The majority of raw materials and components (accessories, vehicle parts and engines) purchased by BRP in the manufacture of vehicles are steel, aluminum, fiberglass, paint, tires and plastics (including thermoplastic and silicone).[14,15]

8 Design

8.1 Product Research and Development

> BRP relies heavily on research and development to maintain the high performance reputation of its products, build customer loyalty and reduce production costs. The company invested C$198.6 million in 2018, representing approximately 4% of its annual sales.[16]

The main concerns of the industry are currently to develop safer and more efficient vehicles in terms of fuel consumption while maintaining the performance standards required by customers.[17] Investments have made it possible to develop products such as Rotax engines, which are found on most BRP vehicles, including ATVs and SSVs. These engines are known for their high performance, but also for their fuel economy and lower GHG emissions.

Other efforts have also been made in the choice of materials, with the use of high-strength steel to reduce the weight of Can-Am vehicles and thus reduce fuel consumption.[18]

Within the Marine Propulsion Systems (MPS) division, BRP meets with all strategic suppliers twice a year to ensure the alignment of critical business objectives. For example, divisions meet with their suppliers to encourage innovation in many areas such as lightweight materials, fuel consumption, alternative fuels and overall emission reductions for products or operations.[19]

---

[13] Ibid.

[14] Interview with Mr. Gagnon, General manager of the Fédération Québécoise des Clubs Quads, November 29, 2018.

[15] Ediz Ozelkan (2018), op. cit.

[16] BRP Inc., (2018) "Annual information form: fiscal year ended January 31, 2018".

[17] Ediz Ozelkan (2018), op. cit.

[18] BRP Inc., "Corporate Social Responibility—Fiscal Year 2018", op. cit.

[19] BRP Inc.,"Latest data of BRP Corporate Social Responsability", https://brp.metrio.net/, April 5, 2019.

8.2 *Product Policy*

BRP's research and development activities aim to innovate and launch new products quickly and regularly. In 2018, the company owned 59 SSV models and planed to market a new one every six months until 2020.[20]

9 Production

BRP vehicles are manufactured at the request of customers through dealerships and during annual presentations of new product lines made in the various clubs.[21] Generally, dealers pre-order the basic models in order to have products immediately available.[22]

Each BRP plant determines its own priorities in terms of energy consumption in order to respect its production context, but also the laws and standards of the country where it is located. In 2012, for example, a European directive was adopted imposing measures to improve the European Union's energy efficiency by 20% by 2020. BRP's European plants have therefore implemented various measures to achieve this objective.

Energy consumption for BRP production worldwide has been increasing steadily since 2014. The main sources are electricity, natural gas and oil.

9.1 *Environmental Management of Production*

In order to limit the waste generated by the plants, BRP melts and re-moulds certain defective plastic parts into new parts. At the same time, BRP is assessing the relevance of applying ISO 14,001 to all its plants, although production sites are developing their own environmental management systems.[23]

The initiatives of the plants are diverse. For example, the Sturtevant plant in the United States has been awarded the title of "Green Master" by the Wisconsin Sustainable Business Council for the development of a wastewater system. This system allows the site to reuse water from outboard engine tests for other processes in the plant, saving more than 83 million liters of water and US$175,000. The Querétaro plant in Mexico has taken the initiative to optimize its recycling processes by creating a sorting area for all operational and office waste. This makes it possible to sort the

[20] BRP Inc., (2018) "Annual information form: fiscal year ended January 31, 2018".

[21] Les clubs sont des regroupements de personnes pratiquant une même activité dans une certaine région (ex.: le club de motoneigistes «Les sentiers Rocher-Percé» de la région de la Gaspésie au Québec).

[22] BRP Inc., "Corporate Social Responibility—Fiscal Year 2018", op. cit.

[23] Ibid.

waste on site, compact it and sell it to a waste recovery company, thus reducing travel. The Gunskirchen plant has an electricity recovery system using regenerative brakes.[24]

10 Logistics/Production Optimization

The majority of BRP vehicle accessories and parts are manufactured, designed and produced in-house. However, for some components, the company must use external suppliers. The latter are audited and evaluated according to their quality management system, their ability to meet delivery deadlines, their price competitiveness, their innovation/technology and their compliance with laws and CSR (corporate social responsibility) standards. Audits and evaluations are conducted annually to minimize risks and improve practices.[25]

One of BRP's strategic priorities is to improve business processes. The company therefore tries to optimize its processes by considering several factors such as "proximity to key retail markets, the presence and cost of skilled labor, production capacity, international and local laws, rules and regulations (including tariff and duty agreements and free trade), and social and political conditions".[26]These various factors are analyzed regularly in order to adapt the manufacturing strategy.

BRP therefore plans the production of parts and vehicles according to the distance from the target market. For example, Lynx snowmobiles for the European market are produced in factories in Europe, as are their Rotax engines, in order to avoid production and delivery from America.[27] However, the choice of the location of some plants is also motivated by a reduction in labor costs, such as the opening of the Querétaro in 2013 and Juárez 2 in 2014 plants in Mexico.[28]

11 Distribution

In terms of supply and distribution networks, BRP is present throughout the world. With its suppliers in Quebec and Mexico, it has developed reusable boxes and containers to reduce packaging waste and the weight of deliveries, and to optimize the use of space. The Querétaro plant manages to achieve 90% space occupancy in its shipments to the Valcourt plant.[29]

---

[24] BRP Inc.,"Latest data of BRP Corporate Social Responsability", op. cit.

[25] BRP Inc.,"Latest data of BRP Corporate Social Responsability", op. cit.

[26] "BRP INC.—Notice annuelle 20 mars 2018"

[27] Ibid.

[28] Ediz Ozelkan (2018), op. cit.

[29] BRP Inc., "Latest data of BRP Corporate Social Responsability", op. cit.

Vehicles are transported from international distribution centers to dealers through contractual carriers that allow faster delivery, or directly by BRP. In 2018, the company was selling directly to 3,200 dealers in 21 countries and through a network of 185 distributors to sell to 915 other dealers in more than 79 countries.[30]

BRP generally enters into a contract with its dealers to allow them to sell certain products; in exchange, dealers keep a certain inventory of spare parts to provide warranty and non-guarantee repair service to users of BRP vehicles.

12 Customers and Uses

Most of recreational vehicles are purchased by consumers through a dealer (54% in Canada[31] and 77% in the United States[32]). This sector includes vehicles used for activities such as hiking, hunting, running, etc. Other buyers are exporters, tourist complexes, parks and golf clubs, as well as farmers or other workers.[33,34]

12.1 Snowmobile

In 2018, the average age of snowmobilers in Canada was 45 years. Snowmobiles have two uses:

1. Family and recreational (95% of uses), allowing users to enjoy landscapes that are usually inaccessible, to spend time with friends or family and to get closer to nature.[35]
2. Transport vehicle to more remote areas of the United States and Canada for scientific research in the forest and rescue operations.[36]

In 2018, there were approximately 600,000 snowmobiles registered in Canada and 1.2 million in the United States. Snowmobilers in North America travelled an average of 2,012 km per year.

Mr. Garneau, of the FCMQ (Fédération des Clubs de Motoneigistes du Québec), explains the decline in global snowmobile sales in recent years due to the growing urbanization of the population. This increases the cost of doing the activity for

[30] "BRP INC.—Notice annuelle 20 mars 2018"

[31] Chiffre qui inclut les ventes de motoneiges, véhicules tout terrain et de voitures de golf.

[32] Chiffre qui inclut les ventes de véhicules tout terrain seulement.

[33] Ediz Ozelkan (2018), op. cit.

[34] Devin Savaskan, (2018), op. cit.

[35] "Snowmobiling Fact Book", http://snowmobile.org/docs/isma-snowmobiling-fact-book.pdf. November 20, 2018.

[36] Ibid.

individuals living in the city who have to store their vehicles in a garage. In addition, he says winters are more capricious and unpredictable than before, making snowmobiling more difficult in some parts of Quebec and the rest of Canada.[37]

13 ATV and SSV

Most ATV or SSV (Quad) users in Canada are baby boomers, individuals born between 1946 and 1964. In the United States, the average age of the majority of Quad vehicle owners is around 40 years old.[38] The great popularity of SSV over ATVs is due in part to the aging of users, who prefer to use slower and safer recreational vehicles than snowmobiles, considered as a more sporty activity. ATV and SSV prices are also a barrier for young adults under 35 and young families, while many people remain curious and would like to practice them from time to time or simply try.[39]

Quad is therefore an activity favored by families and older people. In addition, these are vehicles that can be used 12 months a year, unlike snowmobiles, which are only used for a few months a year. ATVs are considered less comfortable than SSVs because people sit as if on a motorcycle, one behind the other. By contrast, people sit side by side in a SSV like in a car. Quad riders mainly practice this activity to drive around, discover picturesque circuits and new destinations.[40]

In Canada, most off-road vehicle sales come from the provinces of Ontario and Quebec.[41] In the United States, California leads the market, mainly because of the presence of many recreational facilities and programs that encourage youth to participate in this activity.[42]

---

[37] Interview with Mr. Garneau, manager external relations and strategic development of the Fédération des Clubs de Motoneigistes du Québec, November 28, 2018.

[38] "Share of Americans Who Own An Atv (All-terrain Vehicle) in 2018, by Age.", www.statista.com/statistics/228874/people-living-in-households-that-own-an-atv-all-terrain-vehicle-usa/, April 12 2019.

[39] Interview with Mr Gagnon, op. cit.

[40] Ibid.

[41] "MOTORCYCLE, SCOOTER & OFF-HIGHWAY VEHICLE ANNUAL INDUSTRY STATISTICS REPORT" (2017), https://www.cohv.ca/wp-content/uploads/2014/11/Annual-Motorcycle-and-Off-Highway-Vehicle-Industry-Report-from-MMIC-and-COHV-2017.pdf. November 20, 2018.

[42] Amulya Agarwal and Akshay Prakash, (2019) "U.S. Off-Road Vehicles Market Growth 2018–2024 Industry Share Report", https://www.gminsights.com/industry-analysis/us-off-road-vehicles-market. April 5, 2019.

13.1 Clienteles and Demographic Trends

Most BRP's US customers are 40–50 and over and is aging.

Young Canadians aged 15 to 35 are different from previous generations; they are more diverse, connected to new technologies, socially engaged and educated. On average, in 2016, over 85% of young Canadians lived in a large urban areas and 27% were members of a visible minority, as compared to just over 10% among those aged 65 and over. Young Canadian adults aged 20 to 34 increasingly live with their parents, 35% of them are living at home. In the United States, 1 in 3 adults aged 18 to 34 live with their parents, or about 24 million people.[43] This trend can be explained by a lower proportion of young people starting families early and rising real estate costs.[44] In addition, young Americans are increasingly educated, as well as with significant debts. In 2013, 41% of young families had student debt, compared to 17% in 1989.[45]

However, consumers also have a wide choice of vehicles offered at attractive prices by various manufacturers. This changing industry forced Canadian companies to develop their comparative advantages in research and development to differentiate themselves from foreign companies that rely on low prices.[46]

13.2 Acceptance of Recreational Vehicles in Canada

In general, Canadians are more accepting of snowmobiling than they are of ATVs or SSVs. Indeed, 37% of respondents do not say they are very or not at all in favor of snowmobiling, compared to 45% for ATVs and SSVs.[47]

The difficult acceptance of these vehicles is mainly due to their environmental impacts on pollution, damage to fauna and flora as well as to the high risk of accidents. In 2016 and 2017, 2,834 ATV and SSV accidents, and 911 snowmobile accidents were reported in Canada.[48] In the United States, the number of injuries recorded

---

[43] Jonathan Vespa, (2017) "The Changing Economics and Demographics of Young Adulthood: 1975–2016" https://www.census.gov/content/dam/Census/library/publications/2017/demo/p20-579.pdf. April 10, 2019.

[44] "Un portrait des jeunes Canadiens", (2018) https://www150.statcan.gc.ca/n1/pub/11-631-x/11-631-x2018001-fra.htm. April 10, 2019.

[45] Jonathan Vespa, (2017), op. cit.

[46] Ediz Ozelkan (2018), op. cit.

[47] "Based on Your Experience with These Sports and Activities, or Just Your General Impression, How Favourable Are You towards The Following?.", www.statista.com/statistics/472440/general-opinion-on-power-sports-industry-in-canada-by-category/, April 8, 2019.

[48] Leslie young, (2018) "Top 5 sports and activities that land Canadians in the hospital", https://globalnews.ca/news/4316749/sports-injuries-hospitalizations-canada/. April 8, 2019.

from off-road vehicle use was 101,200 in 2016.[49] In all, customers mainly demand high-performance and safe products.

14 Maintenance and Repair

Motorized sports vehicles, being long-lasting products, can easily be repaired by a specialist (garage or dealer) or by its owner if he/ she has the necessary knowledge in mechanics.

New BRP products purchased from a dealer carry a minimum six-month warranty for ATVs in Canada and the United States, as compared to two years in Europe.[50] When relatively new used models are purchased from a dealer, the initial purchase warranty may still be valid for the second owner. Otherwise, for vehicles that no longer have a BRP warranty, a minimum consumer warranty of 30 days and 500 km is applicable in Quebec, Canada, for example, but may vary in other countries.[51] It is also possible to obtain longer guarantees by taking advantage of promotions or by paying a surcharge to obtain B.E.S.T. coverage. This additional coverage extends the vehicle's coverage by 12, 24 or 36 months, with no mileage limit. It covers vehicles throughout North America, which means that repairs can be done at any authorized BRP dealers in Canada, United States and Mexico.[52]

There are differences in repair between product families. Some models, such as snowmobiles or personal watercraft, are more complex to repair than before because of new technologies that require some expertise from the repairer. By contrast, ATVs and SSVs are much easier to maintain because their mechanics are simpler. However, most of the maintenance of BRP vehicles is done by dealers rather than by the users themselves.[53]

Complex mechanics also generate higher maintenance costs for snowmobiles than for ATVs and SSVs and influence vehicle life. Quads, which are on wheels, generally have a longer lifespan than snowmobiles, which advance with a friction system, because the mechanics force less and damage is less rapid.[54] The average annual cost of operating an ATV or a SSA in Quebec is C$700 compared to C$2,000 for a snowmobile. This price includes trail access permit and maintenance.[55]

[49] John Topping, (2017) "2016 Annual Report of ATV-Related Deaths and Injuries" https://www.cpsc.gov/s3fs-public/atv_annual_Report_2016.pdf?vIcLfTM9VNDc23qe6FQyhJq7A7454xCr, April 10, 2019.

[50] BRP Inc., "BRP Best Potection" (2013) https://can-am.brp.com/content/dam/canam-offroad/United-States/English/MY2014/Documents/BESTwarranty/BRP_BestUSA_Dep.pdf, April 10, 2019.

[51] Interview with a BRP distributor, April 10, 2019.

[52] BRP Inc., "Service prolongé B.E.S.T.", https://www.brp.ca/sur-route/proprietaires/best-extended-service.html, April 14, 2019.

[53] Interview with Mr Gagnon, op. cit.

[54] *Ibid.*

[55] *Ibid.*

15 Second Life

Once owners decide to sell their vehicle, they have a choice between: selling it on the used market or exchanging it for a newer model at some dealerships and paying the difference. Dealers who take back BRP vehicles can recondition and resell them, but they usually do not offer any additional warranty over and above that given at the time of purchase of the new product. Dealers generally decide not to take too old models in order to avoid mechanical problems and to satisfy demand in terms of performance and aesthetics.[56]

Most recreational vehicle users use both options, as they change vehicles on average every two years for snowmobiles and every five years for ATVs and SSVs.,[57,58] There is also a relatively large market for used motor vehicles on sales sites between private individuals such as Kijiji, or directly on dealers' websites.

Renting is a business model that is now common in many industries (tooling, automotive, real estate, etc.). There are some organizations that rent recreational vehicles mainly for regional tourism, but their offer is expensive. Some specialized rental companies only offer rental, without transporting the vehicle to the place of leisure. The customer must therefore transport it himself to the place where the motor sport is practiced. Prices for the rental of a basic Quad type vehicle start at C$139 plus taxes for 4 h, and at C$350 plus taxes for a weekend. SSVs are usually more expensive because they offer more comfort to users and sometimes allow more than two people to be transported. These rental companies hold several models of BRP Can-Am ATVs and SSVs. Customers can therefore rent and test BRP vehicles.

16 End of Life

The service life of off-road vehicles depends to a very large extent on the type of engine, maintenance performed and use by its drivers. In addition, some vehicles such as snowmobiles are used only a few months of the year, while ATVs and SSVs can be used all year round.

Owners usually dispose of vehicles once repair costs exceed the resale value of the vehicle. The sale of the vehicle in spare parts remains a minority. Most vehicles end their lives at the landfill or in backyards, garages or barns.[59]

Sherbrooke, 2018. The BRP team notes that much has been accomplished in terms of sustainable development by the company, but it does not know which circular economy strategies would be most relevant. The company wondered what needed to be put in place at the organizational level to improve its environmental performance and become a sustainability leader in the powersports industry.

---

[56] Interview with a BRP distributor, op. cit.

[57] Interview with Mr. Gagnon, op. cit.

[58] Interview with Mr. Garneau, op. cit.

[59] Interview with Mr. Gagnon, op. cit.

Appendix 1 The Circular Economy

Circular economy is "a production, exchange and consumption system that optimizes the use of resources at all stages of the life cycle of a good or a service, in a circular logic, while reducing the environmental footprint and contributing to the well-being of individuals and communities" (Quebec circulaire 2021).

Circular economy strategies	
Ecodesign	Ecodesign aimsto take minimize the potential environmental impacts of products. Circular economy focuses on the optimal use of resources. In particular, designers will be invited to: Design products that simultaneously meet several functions Reduce the amount of resources required to manufacture and use a product Favor low-impact resources (renewable, non-toxic, reusable, recycled, etc.) Promote prolonged use of the product (durable, repairable, easy updates) Be inspired by nature in product design (biomimicry). Nature does not know waste and benefits from billions of years of evolution
Responsible consumption	The circular economy model leads to the identification of new procurement criteria focused on the optimal use of our resources. Individual and organizational consumers obviously have a key role to play in reducing resource consumption At the organizational level, the development of new business models, such as the economy of functionality and the sale of used or refurbished products, will encourage some companies to switch to closed-loop logistics. They will be able to source their products at the end of their life cycle or from similar products. This approach allows for a better recovery rate and good control over the quality and quantity of material available for the manufacture of new products To learn more about responsible consumption, visit the following site: ECPAR: https://www.ecpar.org/en
Operations optimization	Throughout the value chain, from raw material extraction to product distribution, companies' operations are based on the consumption of natural resources (water, energy, metals, etc.). Motivated by the reduction of procurement costs or by environmental and good management approaches (e. g. ISO 14,001, Lean), many of them have already begun a shift in resource consumption reduction It is also possible to take advantage of new technologies, in particular by: Information management systems to better manage resource consumption, target losses, and effectively plan distribution logistics Additive manufacturing (including 3D printing) also has interesting potential for saving resources

(continued)

(continued)

Circular economy strategies	
Collaborative economy	The collaborative economy encompasses a wide variety of business strategies and trade models to maximize the use of goods and products in circulation in the market. From citizen initiatives to commercial platforms such as UBER and AirBNB, the collaborative economy is booming internationally
Renting	In rental, an organization or individual owns a property and leases its use for a fixed period of time. This strategy is well known in some sectors, such as tooling, automotive and real estate. This business model therefore meets the same objective as the collaborative economy, but is more suitable for clients who do not wish to be in contact with other users of the same good In leasing, the management of end-of-life products may be less efficient than in functional economy where the manufacturer retains ownership and can therefore more easily recover, repair and recondition them. However, landlords also have an interest in ensuring proper maintenance and repair of assets in order to maximize their return on investment
Maintenance and repair	Maintenance and repair can be carried out by the consumer himself, a specialized organization (e. g. shoemaking), the distributor or the manufacturer Today, the cost of repair is often higher than the purchase of a new similar product.. The attractiveness of the new also hinders repair and the increasingly rapid technological progress means that the products to be repaired are already completely obsolete after only two or three years of use. Fortunately, consumer mentality is changing and several interesting opportunities are now available, including: The iFixit.com website, which contains thousands of free repair guides and sells parts online Repair Café, which are ephemeral repair workshops where residents can bring their equipment for repair and be accompanied by experts
Donating and reselling	Donation and resale allow consumers or organizations to put back into circulation products that they no longer need, but which are still in good condition. These strategies are far from recent, but new digital platforms offer unprecedented opportunities to connect those who want to dispose of objects with those who are looking for them The Kijijiji 2017 Index shows that: Among the products most frequently redistributed are baby items, games and clothing Purchasing goods at a lower price is the main source of motivation to acquire second-hand goods 2017 Kijiji index report: https://www.kijiji.ca/kijijicentral/app/uploads/2016/08/2017-Index1.pdf

(continued)

(continued)

Circular economy strategies	
Refurbishing	Refurbishment consists in restoring a product or component to new condition with a guarantee equivalent or close to that of new. The product is collected, transported, disassembled, each of its components is cleaned and controlled, some of them changed or remanufactured. The product is then reassembled, controlled and put back on the market. The possibilities for reconditioning depend a lot on the industrial sectors. In the transport sector, heavy or military equipment, with very long life cycles (boats, trains, planes, helicopters), is mainly reconditioned, but less so for lighter finished products, such as automobiles To facilitate the deployment of this strategy, products must be designed taking into account the constraints of their end-of-life treatment
Performance economy	This new business model does not focus on selling products to consumers and buyers, but rather on selling the use of these same products. Users therefore buy the function and not the product itself. For example, Michelin offers owners of heavy truck fleets a tire service rather than the tires themselves. The customer is billed per kilometer and does not own the tires By retaining ownership of the product, the manufacturer can adequately manage the end of the cycle. The product can then be repaired, reconditioned or dismantled to generate new components or raw materials. The manufacturer is therefore freeing itself, in part, from the volatility of commodity prices Performance economy explained in video: https://www.youtube.com/watch?v=Cd_isKtGaf8
Industrial ecology	Industrial ecology aims to optimize the use of resources by industrial companies in a territory by drawing inspiration from the cycles of natural ecosystems. It sets up exchanges (synergies) of material; energy or resource flows between two or more companies. The development of a by-product, waste, or available energy, thus allowing an environmental gain while presenting new business opportunities for symbiotic companies. A distinction is made between: Substitution synergies, where the residue of one replaces all or part of a raw material of the other; Pooling synergies, where several companies coordinate their resource needs

(continued)

(continued)

Circular economy strategies	
Recycling and composting	Recycling: Recycling is the use, in a manufacturing process, of a recovered material to replace a virgin material. The circular economy makes it possible, on the one hand, to set up the shortest possible recycling loops and thus to favour local recycling markets over export markets. On the other hand, in order to preserve the value of resources, it invites to focus on recycling in products with high added value (overcycling) Composting: In a circular economy, organic matter returns to the soil to enrich it. In this sense, composting is an aerobic treatment (in the presence of oxygen) of organic matter, which creates a mature solid product: compost
Energy recovery	Recovery is any operation that does not constitute disposal and that aims to obtain useful products or energy from residual materials Energy recovery, for its part, can involve thermal treatment processes that will irreparably transform materials such as: incineration with energy recovery, combustion in an industrial boiler or cement kiln, pyrolysis and gasification. According to the Environment Quality Act, the thermal destruction of residual materials constitutes energy recovery as long as this treatment of the materials meets the regulatory standards prescribed by the government, including a positive energy balance and the minimum required energy efficiency, and contributes to the reduction of greenhouse gas emissions. T

Source authors from https://www.quebeccirculaire.org/index,en.html

References

1. Agarwal A, Prakash A (2019) U.S. Off-Road Vehicles Market Growth 2018–2024 Industry Share Report. https://www.gminsights.com/industry-analysis/us-off-road-vehicles-market. Accessed 5 Apr 2019
2. BRP Inc. (2018) "Corporate Social Responibility – Fiscal Year 2018", https://www.brp.com/content/dam/corpo/Global/Documents/csrdocuments/BRP_Overview%20GRI2018_EN_final.pdf. Accessed 13 Apr 2018
3. BRP Inc. Plus de 75 ans d'innovation. https://www.brp.com/fr/a-propos-de-brp/our-story.html. Accessed 17 Nov 2018
4. BRP Inc. (2018) Annual information form: fiscal year ended January 31, 2018. http://2017.brp.com/fr/responsabilite-sociale-d-entreprise/
5. BRP Inc. (2017) "Annual information form: fiscal year ended January 31, 2017
6. BRP Inc. (2016) Annual information form: fiscal year ended January 31, 2016
7. BRP Inc. (2015) Annual information form: fiscal year ended January 31, 2015
8. BRP Inc. (2014) Annual information form: fiscal year ended January 31, 2014
9. BRP Inc. (2018) Corporate Social Responibility – Fiscal Year 2018. https://www.brp.com/content/dam/corpo/Global/Documents/csrdocuments/BRP_Overview%20GRI2018_EN_final.pdf. Accessed 19 Nov 2018

10. BRP Inc. (2013) BRP Best Potection https://can-am.brp.com/content/dam/canam-offroad/United-States/English/MY2014/Documents/BESTwarranty/BRP_BestUSA_Dep.pdf. Accessed 10 Apr 2019
11. BRP Inc. Service prolongé B.E.S.T. https://www.brp.ca/sur-route/proprietaires/best-extended-service.html. Accessed 14 Apr 2019
12. BRP Inc. Latest data of BRP Corporate Social Responsability. https://brp.metrio.net/. Accessed 5 Apr 2019
13. Savaskan D (2018) ATV Manufacturing in the US. IBISWorld Industry Report
14. Ozelkan E (2018) ATV, Golf Cart & Snowmobile Manufacturing in Canada. IBISWorld Industry Report
15. Ipsos. Based on Your Experience with These Sports and Activities, or Just Your General Impression, How Favourable Are You towards The Following? www.statista.com/statistics/472440/general-opinion-on-power-sports-industry-in-canada-by-category/. Accessed 8 Apr 2019
16. McQuarrie J (2006) Bombardier Inc. https://www.thecanadianencyclopedia.ca/fr/article/bombardier-inc. Accessed 17 Nov 2018
17. Vespa J (2017) The Changing Economics and Demographics of Young Adulthood: 1975–2016. https://www.census.gov/content/dam/Census/library/publications/2017/demo/p20-579.pdf. Accessed 10 Apr 2019
18. Topping J (2017) 2016 Annual Report of ATV-Related Deaths and Injuries. https://www.cpsc.gov/s3fs-public/atv_annual_Report_2016.pdf?vIcLfTM9VNDc23qe6FQyhJq7A7454xCr. Accessed 10 Apr 2019
19. Yyoung L (2018) Top 5 sports and activities that land Canadians in the hospital. https://globalnews.ca/news/4316749/sports-injuries-hospitalizations-canada/. Accessed 8 Apr 2019
20. Motorcycle and moped industry council (2017) MOTORCYCLE, SCOOTER & OFF-HIGHWAY VEHICLE ANNUAL INDUSTRY STATISTICS REPORT. https://www.cohv.ca/wp-content/uploads/2014/11/Annual-Motorcycle-and-Off-Highway-Vehicle-Industry-Report-from-MMIC-and-COHV-2017.pdf. Accessed 20 Nov 2018
21. Powersports Business. Projected Global All-terrain Vehicle Market Share in 2018, by Key Manufacturer. www.statista.com/statistics/438085/global-all-terrain-vehicle-market-share/. Accessed 3 Apr 2019
22. Powersports Business. Projected Global Side-by-side All-terrain Vehicle Market Share in 2018, by Vendor. www.statista.com/statistics/438113/global-side-by-side-atv-market-share/. Accessed 3 Apr 2019
23. Québec Circulaire, "Stratégie de circularité", https://www.quebeccirculaire.org/static/strategies-de-circularite.html. Accessed 4 Apr 2019
24. Snowmobile. Snowmobiling Fact Book. http://snowmobile.org/docs/isma-snowmobiling-fact-book.pdf. Accessed 20 Nov 2018
25. Statista Survey. Share of Americans Who Own An Atv (All-terrain Vehicle) in 2018, by Age. www.statista.com/statistics/228874/people-living-in-households-that-own-an-atv-all-terrain-vehicle-usa/. Accessed 12 Apr 2019
26. Statistique Canada (2018) Un portrait des jeunes Canadiens. https://www150.statcan.gc.ca/n1/pub/11-631-x/11-631-x2018001-fra.htm. Accessed 10 Apr 2019

Fashion Upcycling: A Canadian Perspective

Jennifer L. Dares

Abstract Upcycling is a design practice that uses pre- and post-consumer textile waste derived from apparel manufacturers or disassembled garments to create new fashion, providing a sustainable design solution to divert textile waste from landfills. The qualitative research study underpinning this chapter examined challenges faced by Canadian designers who upcycle and strategic solutions they integrated into their business models. The study used purposive sampling to recruit participants who had more than 10 years of textile experience and had produced an upcycled retail collection for at least 2 years. Participants included representatives from Brand A, whose accessories have sold in 20 stores (including Simon's in Montreal); Brand B, featured in *British Vogue*, with stores in the U.K. and Sweden; Brand C, a children's wear brand that received accolades from fashion icon Jeanne Beker; Brand D, a Canadian upcycling pioneer with collections retailed at Anthropology, Holt Renfrew, Roots, and Sporting Life; and Brand E, the winner of the 2018 CAFA Fashion Impact and the H&M Sustainability awards. Interview findings revealed design challenges related to on-trend, ever-evolving fashion styles; achieving high volume; and training design teams to recognize "good" versus "bad" damage through the brands' respective aesthetic lens. Main production challenges involved scale, labour, and cleaning, while retail challenges included price, stigma, and narrative. Findings reported in this chapter contribute to knowledge regarding strategies for designing, producing, and retailing upcycled fashion both in Canada and globally. This study is important to academic community members who wish to engage in upcycling, and offers tangible strategies to participate in a circular economy, divert textile waste from landfills, and potentially alleviate climate change.

Keywords Fashion · Design · Sustainability · Upcycling · Redesign · Textile waste · Sustainable clothing · Production

J. L. Dares (✉)
Toronto, Canada
e-mail: jennifer.dares@senecacollege.ca

S. S. Muthu (ed.), *Circular Economy*, Environmental Footprints and Eco-design of Products and Processes, https://doi.org/10.1007/978-981-16-3698-1_2

1 Introduction

Textile waste is an increasingly complex and wicked problem [40]. The Environmental Protection Agency reported that in 2013, 15.1 million tons of textile waste was generated, of which 12.8 million tons were discarded [15]. While there is a lack of Canadian data concerning textile waste, U.S. textile waste is a growing concern, with 11.3 million tons were deposited into landfills in 2009, representing a 40% increase since 1999 [61]. CBC Marketplace, a Canadian media outlet, also featured the problems surrounding textile-waste in North America [14]. The *Pulse of the Fashion Industry 2019 Update* reported that although many in the fashion space have been working toward implementing more sustainable practices, they have not done so at the rate needed to reach the 2030 Sustainable Development Goals, a division of the United Nations [26]. Since this research study was undertaken in 2018, the numbers of town and city councils that have declared a climate emergency have increased [13].

The fashion industry and academic community have been called upon to generate solutions to this problem and to develop recyclable fibres for textiles; however, this may take some time and scalable solutions are not viable yet. Therefore, it is essential to implement interim solutions to the textile-waste problem [19, 27]. Fashion upcycling is a design strategy that holds promise as a proposed solution to textile waste [19, 28, 47, 62]. Fashion upcycling is a design practice that uses post-consumer or pre-consumer textile waste to create new clothing and accessories, therefore providing textiles with a second life and diverting them from landfills [5, 19, 25, 27, 28, 31, 59, 62]. The upcycling redesign could involve minimal modifications such as adding embellishment to more labour-intensive processes that involve the complete deconstruction and reconstruction of a garment [30]. There is much speculation that is critical of or supports upcycling as a sustainable design solution,however, there is scant academic research on this topic and past scholars have suggested further research is needed [11, 47, 62].

This chapter investigates the successes and challenges of creating upcycled apparel in Canada, taking into consideration the design and manufacturing processes as well as retailing systems. The inquiry began with a literature review. Following this, Canadian designers working in the realm of upcycling were recruited to complete a demographic survey and participate in semi-structured interviews. The designer's online presence was also analyzed to triangulate the interview findings. The importance of this study is relevant to educating designers, students in postsecondary, fashion and retailing programs as well as entrepreneurs and retailers so that they might better understand and embrace fashion upcycling. An increase in fashion upcycling is a viable solution to accumulating waste as it would divert more textiles from landfills.

The study underpinning this chapter examined fashion upcycling through the lens of the designer. The study sought to shed light on the practice of fashion upcycling and its associated challenges, successes, and solutions. Study participants included Canadian fashion companies who have a portion of their company within the Quebec

City–Windsor corridor region. This geographic region was selected due to it being the densest population with the most economic growth in Canada since 1972 [52]. This largely populated area would most likely see more textile waste going to landfill. The research question was: What are the successes, challenges, and solutions of creating upcycled apparel or accessories from the designer's perspective within the phases of design, production, and in the retailing?

According to the United Nations Fashion Alliance, those who are leaders in the $2.4 trillion fashion space must work toward meeting the 2030 Sustainable Development Goals to alleviate the environmental impacts [58]. The amount of textile waste sent to landfills has significantly increased since the beginning of this century [61]. Much of the textile waste is a result of the output of the fashion space in the design, production, and retail of clothing and accessories. People will continue to need and desire new clothes, so alternative practices are needed to divert textile waste from landfill.

2 Chapter Outline

The remainder of this chapter is structured into four sections, encompassing a literature review, methodology, results, and conclusions. The literature review explores research related to textile waste within the fashion space, and the practice of upcycling as a solution toward diverting textile waste from landfills. The methodology provides an outline of the research design, including the methods used, participant recruitment, data collection, data analysis, ethical considerations, validity, and research limitations. The results section includes charts that summarize findings according to challenges and solutions within the phases of design, production, and retail. The conclusion includes a discussion with a particular focus on the contributions of this research study to education, as well as recommendations for further research and the professional practice of fashion design.

3 Literature Review

The research on textile waste reveals that most is going to landfills, causing adverse environmental impacts [1, 6, 11, 28, 39–41, 47]. Scholars assign blame on the many businesses that operate in the realm of fast fashion, along with their insatiable consumers [1, 24, 41, 48]. Proposed solutions centre on fibre recycling, but products need to be designed with recycling in mind. The technology is not currently where it needs to be for large-scale implementation [11, 27, 42, 43, 55, 57].

Over the past decade, scholars have begun exploring fashion upcycling as a viable solution to diverting textile waste from landfills [11, 31, 47, 62] Paras and Curteza [47] note the importance of upcycling and the lack of comprehensive literature. They critically examined 52 academic papers on the phenomena of upcycling, concluding that upcycling is in its infancy and expensive due to labour intensity [47]. Findings indicate that upcycling has potential, but government aid is needed, and the authors recommend that future directions include more empirical studies with fashion designers who upcycle [11, 31, 47, 62]. With this in mind, I explored the successes and challenges encountered by fashion designers engaging in the practice of upcycling. Questions around the role of technology in the success and possibilities of advancing the fashion upcycling practice were included.

Fashion scholars have investigated fashion upcycling in various contexts including post-consumer waste, perceived value, clothing waste, redesign, mass production, standard versus upcycled fashion processes, marker making, consumer upcycling, pre-consumer textile waste, and textile sourcing. Binotto and Payne [5] have studied upcycling in terms of increasing the perceived value via brand narrative by performing a case study with a designer working in Paris. Black [6] conducted multiple case studies to investigate clothing waste and the practice of upcycling or redesign within fashion companies in the U.K and Europe. Cassidy and Han [11] examined the upcycling process and mass production in the U.K. finding that rigorous research is still required to make this a more common practice.

Dissanayake and Sinha [21] studied the remanufacture of discarded garments or textiles into new designs and the processes required during production by interviewing five U.K. companies. They discussed the differences between standard fashion design and upcycled fashion, noting that inspiration is textile driven as opposed to trend driven at the beginning of the design process [21]. Some of the challenges during the production processes were discussed, specifically around disassembly, and cutting [21].

Han et al. [28] conducted a study in the U.K. that analyzed one traditional fashion high street brand's processes compared to six fashion brands that upcycle, and all were within the same market category. Their findings illuminate the differences and similarities between the design, production, and retail processes of standard versus upcycled fashion apparel [28]. Their findings indicated differences during the design process were that upcycled designers were textile driven and standard fashion designers were trend and consumer driven [28]. They discussed some of the challenges during the production processes arising from the various available textiles being used [28]. Their interview results indicated that the designers promoted the brand on social media engaged with community [28].

To advance the upcycling method, some researchers have examined consumer upcycling practices and the contributions that could be made to the fashion space [8, 30, 31, 62]. A few studies have focused on pre-consumer textile waste as a material source for upcycling practices [6, 11, 57]. Cuc and Tripa [20] explored upcycling during the marker making stage as part of the design process to reduce textile waste and increase profits. In [6] *The Sustainable Fashion Handbook*, Hermès has participated in upcycling as they have created a collection label ("Petit h"), in which they utilize the textile waste leftover from markers used when cutting garments for production as well as flawed goods [10, 32]. Muthu [43] studied the upcycling of adult-sized garments transformed into children's wear, ranging from baby to teen.

The way we think about textile sourcing is changing [6, 46] Black [6] suggests that the upcycling technique will be adopted by mainstream designers in the future, moving to a "cradle to cradle" model now referred to as circular or closed loop in the fashion. O'Mahony [46] elaborates on the cradle-to-cradle model, stating that "industrial and post-consumer waste is set to be more widely acknowledged as important resources rather than merely problems to be solved" (p. 307). Black [6] and O'Mahony [46] note the importance of sourcing textiles close to the design and manufacturing locations to reduce the carbon footprint. Textile use and sourcing are key factors in fashion upcycling, solutions in this realm will allow designers to create a new fashion design model that has increased efficiencies.

The literature review revealed that only a small number of the research studies incorporated interviews with fashion design companies who practice upcycling, and the majority of those studies were conducted in the United Kingdom [21, 28, 27, 54]. Although, two U.K.-based studies primarily investigated the processes of fashion upcycling and compared these processes to standard fashion design, there is a gap in research pertaining to the Canadian fashion space and upcycling. Specifically, there is a lack of information regarding the successes, challenges, and solutions experienced by Canadians while upcycling fashion apparel and accessories.

4 Methodology

The central aim of this comparative case study was to better understand the successes, challenges, and solutions experienced by Canadians who practice fashion upcycling [17, 18, 37, 63]. The specific research question guiding the study was: What are the successes, challenges, and solutions of creating upcycled apparel or accessories from the designer's perspective within the phases of design, production and retailing? This section outlines the research methodology adopted for this inquiry, arranged into four sections: (a) research design, (b) data collection, (c) data analysis, and (d) ethical considerations.

4.1 Research Design

This comparative case study research design used qualitative methods [17, 18, 37, 63]. Qualitative methods are well suited to an in-depth look at a phenomenon through a review of the literature and a comparative analysis of a small purposive sample of participants' experiences [17, 35]. A demographic questionnaire was used to gather descriptive statistics about the participants. The three main themes of design, production, and retail were purposeful and deductive and helped to form the research question. Inductive logic was used when analyzing the results and this data was used to formulate the subthemes.

In order to better understand fashion upcycling, I began with a literature review of academic papers and books on the topic of upcycling. I used Google Scholar, Ryerson University Library & Archives, and Seneca Libraries to search peer-reviewed publications. The criteria set for the peer-reviewed articles was that the literature was published no more than 10 years ago due to the increasingly rapid changes within the fashion industry [40]. The literature searched was dated between 2008 and 2018 using the following primary search terms: upcycled fashion, upcycled clothing, upcycling fashion, upcycling in fashion design, redesign, sustainable fashion, and circular economy. With each of the primary search terms, I used the following secondary search terms: clothing, fashion, design, and textiles. This yielded articles most relevant to upcycling from a fashion designer's perspective.

The literature review enhanced understanding of peer-reviewed research and aided in the development of the research question and methodological approach [47] recommend that future research on upcycling take place in the form of empirical inquiry with fashion businesses. Thus, a demographic questionnaire and open-ended interviews with Canadian fashion designers who practice fashion upcycling were completed. The first section of the demographic survey requested participants' preferred email contact, company name and address. The purpose of collecting this personal and identifiable data was to confirm that the location of the business was within the defined geographical area. The email address was required to contact the participant to arrange interview at a time and date that suited them. The demographic survey also inquired about general apparel design experience and specific design experience with upcycled apparel and accessories and lastly, how many employees worked in the company and on the design team.

The primary research strategy included open-ended interviews with Canadian fashion designers who have been producing wholly or partially upcycled apparel or accessory collections that they sold under their fashion design label using a semi-structured interview guide [38]. Inquiry focused on all stages of the upcycling process, from inspiration research to the sourcing of materials, through to the sale of the garment. The first section of the interview guide focused on general questions about the participants' initial interest in fashion, their training, the catalyst for launching an upcycled collection, the history of their brand, and their brand's philosophy. The second section concentrated on design, sourcing, and production, with questions about their design and manufacturing processes, topics included sourcing, trend forecasting, sorting, laundering, time, techniques, fibre content, labelling, care instructions, government guidelines, deterrents, and technology. The third and final section focused on their retailing operations, unpacking information about the brand's narrative, buyer's response, e-commerce, bricks and mortar, consumer response, and the brand's competitors. The final question asks participants: In reflection, what is the biggest success and the biggest challenge when creating an upcycled fashion? Each participant preferred to provide their own photos, therefore I engaged in online research to source the images that were discussed during the interview and when I made my final selections, I emailed each participant to seek approval to use the image, to request a high-resolution copy of the image and to confirm the name of the photographer so that photo credits may be included.

To increase reliability and validity, data triangulation, comparisons, member checking, purposive sampling, and intercoder reliability strategies were used [17]. Validity was developed through intercoder reliability when the transcripts were reviewed by myself, and by my supervisor, Dr. Sandra Tullio-Pow, to confirm the themes and discuss variances. The findings of the demographic questionnaires, the interviews, and the photos (both those researched online and those provided by the designer) were compared to determine where the data overlapped for data triangulation. The photos confirmed what the designers were talking about. Reflexivity was considered during the creation of interview guide to avoid leading questions [37]. Each participant was sent the transcribed interview to verify they said what they meant for member checking [37]. Disadvantages of the methods listed is that all of the collected data is self-reported and therefore may be biased. Triangulation of these methods lends to less risk in the qualitative data analysis [37].

The research limitations of this study were that it was not possible to visit all of the facilities in which each design company conducts its business due to its vast geographical locations. The global lens provided by participants' vast geographic locations outweighs the disadvantage of visits to each of the facilities. The research limitations are the small sample size of this study.

4.2 Data Collection

This section outlines data collection and describes the research setting, sample type, recruitment and selection of participants, a synopsis of data sources, and a summary of the themes focused on during the interviews [35]. Prior to beginning the study, the researcher completed the Tri-Council Policy Statement course on Research Ethics, this aided preparation and submission of an application to obtain approval from the Ryerson University Research Ethics Board (REB) to recruit human participants.

4.2.1 Recruitment and Selection

Fashion apparel and/or accessory designers who have all or a portion of their business located in the Quebec City–Windsor corridor were recruited for this study. The sample type was purposive and non-random, as it was necessary to recruit Canadian participants who had experience within the fashion upcycling category [37]. A criterion sampling method of recruitment was used to select prospective participants. The inclusion criteria required the participants have a minimum 10 years of experience in some facet of the textile industry at any phase in the life of textiles, with a minimum 2 years of experience designing, manufacturing, and retailing upcycled fashion apparel or accessories under their own label. The fashion upcycling practice must have used pre-consumer or post-consumer textile waste. Pre-consumer textile waste is created during the production of the product, and post-consumer textile waste includes items at the end-of-consumer. The participants were also required to have an online presence for their upcycling fashion brand, whether it be a website, Etsy, Instagram, or another form of social media.

The sample included purposive selection of participants who are very knowledgeable in the fashion space and specifically about the topics of pre-consumer textile-waste, post-consumer textile-waste, sustainability, and upcycling. Each of the participant's interview ranged from 40 min up to 1 h and a half, depending on the length of the participant responses. The rich descriptions provided by the participant's provided reliable qualitative data [37].

To search for companies that fulfilled the criteria, snowball sampling was employed along with searches in online magazines, online newspapers, and upcycled fashion brand websites. Social media applications LinkedIn and Instagram were also searched using #upcycledfashion, #upcycling, #upcycled clothing, #torontofashion, #montrealfashion, and #canadianfashion to locate fashion designers with the criteria as mentioned earlier.

The recruitment process began by sending an email message to 12 prospective companies using the contact information listed on the company website or to the using DM (Direct Message) to the company's Instagram account (see Fig. 1). In the recruitment message, the focus of the research was explained, and prospective participants were asked to agree to an interview that would take one and a half hours and complete a short 15-min demographic survey. The recruitment letter offered a $25 gift card from a coffee house of their choice as a small token of appreciation for participating. Those interested in participating in the study contacted the researcher directly via mail. Those individuals who responded were emailed a welcome letter and consent form, allowing time for review in order to pose questions to the researcher

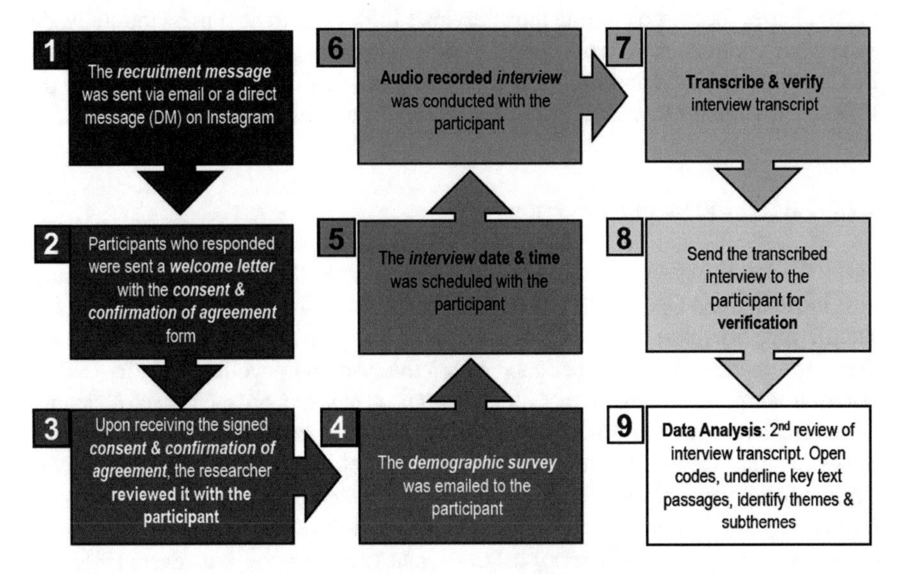

Fig. 1 Data collection procedures chart showing sequence of multiple phases for this study. *Source* Author; graphic design: Harold Madi (2019)

before consenting to participate. Prior to commencing the interview, the researcher again reviewed the consent form to allow time for further clarification. Participants were reminded that they would receive a copy of their interview transcript via email to verify the information and their responses were required within 7 days of receipt. As well, participants were advised that they could withdraw from the study at any point during the interview and any of the data collected up until the time of withdrawal would be deleted.

Limitations encompass designers who do not practice fashion apparel upcycling or fashion accessory upcycling. Designers who upcycle fashion apparel or fashion accessories working in other parts of Canada were not included in this study.

Multiple data sources were used to triangulate results including a demographic questionnaire, transcribed interviews, the brand's online presence, and the photos provided by the designer of each brand. Data collection began with a short demographic survey to provide descriptive statistics of the sample. The demographic survey form was emailed to each participant and returned to the researcher via email. The interview date and time were scheduled with each designer during a time that was convenient for them. Over 300 min of interview audio was transcribed in preparation for the thematic analysis. Photos for the study were provided by the designer.

4.3 Data Analysis

Data collection and analysis were triangulated through multiple sources, the brand's online presence, demographic questionnaires, interviews that were transcribed, memos, and supplementary materials provided by the designer (see Fig. 2). Following each participant's interview, a memo was written, and the audio recording was uploaded to Rev.com for transcription. Upon receiving the transcribed audio, the researcher listened to the audio recording to verify accuracy. Once each transcript was verified by the researcher, it was sent to each participant for verification.

The transcribed interviews were examined and analyzed using thematic analysis with open coding that was developed based on keywords or phrases [16, 37]. A content analysis was done by creating a summary chart for each participant. Each participant's content analysis was done by listing the topics discussed and the frequency of the topic during the interview was documented. During the thematic analysis concepts, themes and subthemes were identified and analyzed for similarities and differences using the constant comparative method [16, 34, 36].

Each of the three main themes—design, production and retail—were assigned a colour and key text was highlighted by colour accordingly. Subthemes were then identified, and a word index chart was created with three columns, the first titled Design, the second titled Production, and the third titled Retail. To ensure clarity a glossary of terms was created based on the language used by the participants as a single word may have multiple meanings depending on who in the fashion space is speaking. A quote index chart was created in Excel with six tabs, one for each of the three themes with a subtheme tab assigned to each category, one for successes

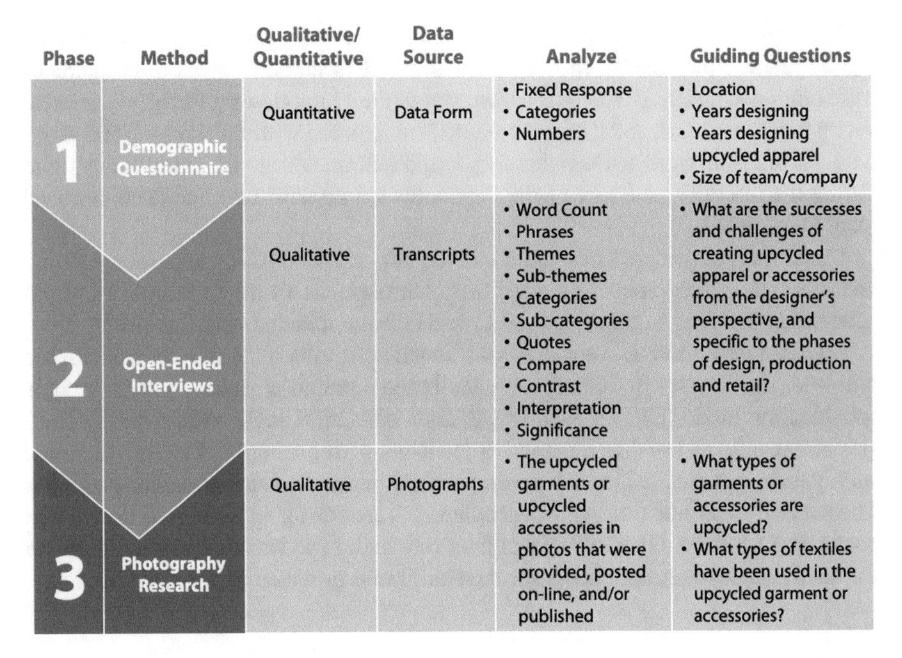

Fig. 2 Data analysis—Summary of methods. *Source* Author; graphic design: Harold Madi (2019)

and a second for challenges and solutions. Contrast and comparison were used to determine the main challenges, solutions, and successes within each theme.

An in-depth thematic analysis was performed on the Index of Themes Chart and the Quote Index Chart that were created using the participants' feedback provided an overview of the challenges, successes, and solutions. During the thematic analysis similarities and differences were noted. A content analysis was performed by documenting the frequency of words or phrases and then looking for similarities or differences amongst the participants. Additionally, the analysis provided the general information required to create a brand comparison chart that was used to develop a descriptive summary of each company's history. To build on past fashion upcycling research the following descriptors were included: the city the company was based in, the year the brand was established, the structure of the company, the market, the product type, the type of textile-waste utilized, where the product was manufactured, where the product was sold, collaborations, community engagement, and the target customer [21, 28] . A chart was created to illustrate the participants' experience in textiles and fashion upcycling and a map was created to illustrate where the participants conducted the various phases of their fashion upcycling business from sourcing textile-waste to retailing.

4.4 Ethical Considerations

The sample type, recruitment, and selection of participants were described in the Data Collection section. The participants were provided with the choice to be assigned a pseudonym for confidentiality, or if they preferred to have their real name and company name published, they were to indicate this on the signatory page of the consent form. If any of the participants chose to be assigned a pseudonym, then all participants would have been assigned a pseudonym. Taking into consideration the significance of speaking with entrepreneurs on the topic of their business practices, it was important at the beginning of each interview when the consent forms were reviewed, to discuss the company name and interviewee name being used. Additionally, member checking was used as each participant was sent the transcribed interview for verification. Although, all participants agreed to be named, and this was done in the brand overview section, they were also assigned a pseudonym of Brand A through Brand E for the purposes of clarity throughout the remainder of the study. The electronic data will be stored until September 17, 2019, upon completion of the research paper and then deleted. The hard copies of all documents will be stored in the office of my supervisor, Dr. Sandra Tullio-Pow for 6 months following the completion of the paper and then all documents will be shredded.

The research data was managed digitally on the Google Drive since it is encrypted and minimizes the data being accessed by unauthorized individuals. Within the researcher's Ryerson Google Drive, a folder labelled MRP was created that contained a numbered folder for each participant, a Quote Chart and an Index of Themes Chart. Within each participant's numbered folder, the following electronic data was stored: the consent forms, photos provided by the participant, a theme chart, the MP4 audio files of the interviews and the transcribed interviews in the form of a Word document. Hard copies of the participant's company name, address, and email, along with the signed consent forms and demographic survey were stored in a locked filing cabinet for 6 months. Physical folders were created for each participant and numbered in the same manner as the digital folders. The physical folders included 11 $\times$ 17-inch hard copies of the transcribed interviews that had been highlighted by theme with notes on the right side and a themes chart. A copy of these documents was provided to the researcher's supervisor so the data could be reviewed for intercoder reliability. These documents were destroyed upon completion of the research paper.

This methodology section reviewed the methods undertaken within this qualitative study. The following section includes a brand overview for each of the participants' companies to provide context the results.

5 Results

The findings of the interviews and the photos provided by the participants are presented and analyzed in the following three sections. First, an overview of each brand is presented to provide context for the results. The following section provides the findings of the qualitative study using thematic coding, a content analysis, and a comparative analysis presented within the three main themes: design, production, and retail. The thematic coding was based on the research question and interview guide's three main themes of design, production and retail. The interview data provided the subthemes developed from the coding and the content analysis. The thematic coding was compared for similarities and differences.

The participants (n = 5) produced wholly or partially upcycled apparel or accessory collections with all or some of the phases of their business within the Quebec City–Windsor corridor, a geographic region in Canada. A map was created to illustrate each brand's geographic locations where the main activities, including design, manufacturing and retail, took place (see Fig. 3). A chart with the main findings of each participant's experience with textiles and fashion upcycling was provided. Each of the five fashion designers had a minimum 10 years of experience with textiles and a minimum of 2 years' experience producing an upcycled fashion collection, sold at retail (see Fig. 4). The knowledge the fashion designers have gained through their experience meant they were able to provide rich perspectives on the state of the upcycling practice in the fashion industry today. Four of the five participants were

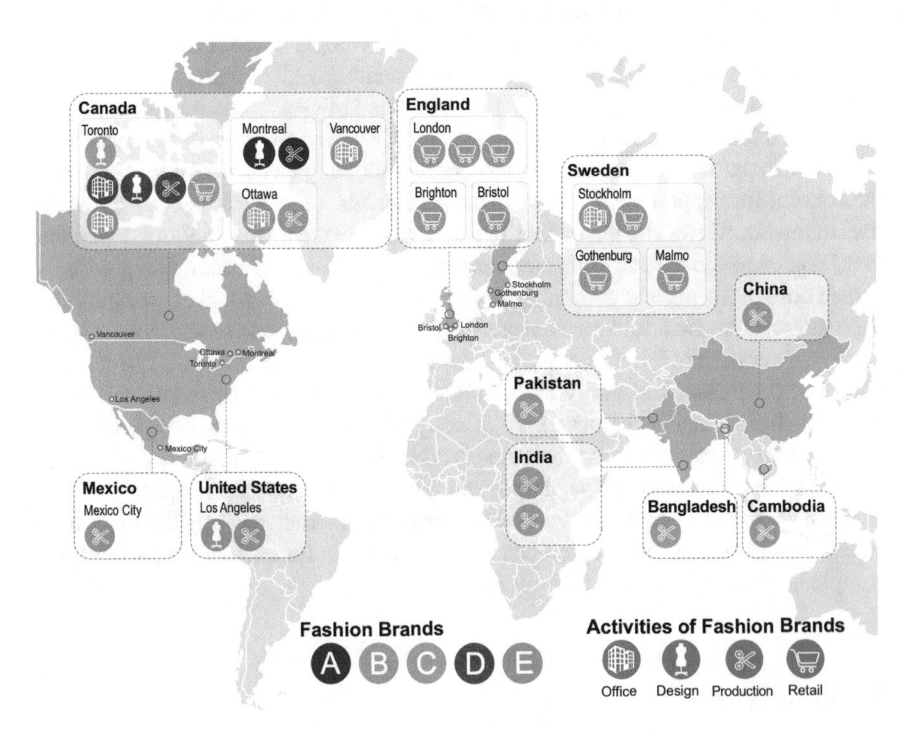

Fig. 3 Geographic locations of each brand's key activities within fashion upcycling. *Source* Author; graphic design: Harold Madi (2019)

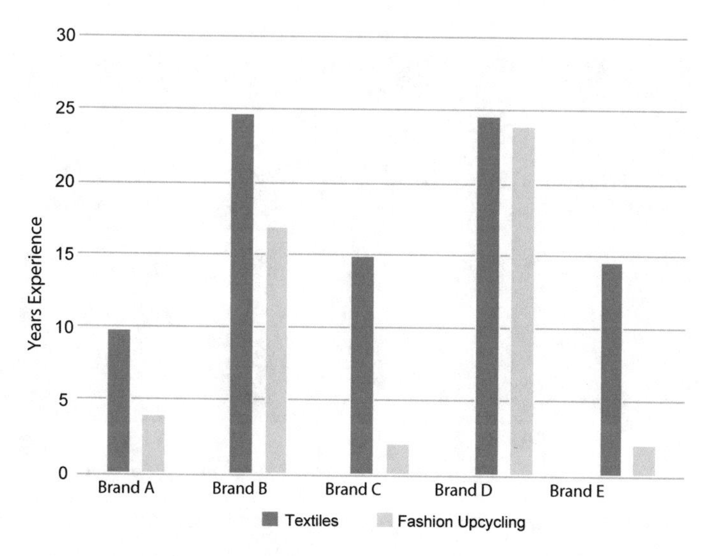

Fig. 4 Main findings of participants' experience with textitles and fashion upcycling. *Source* Author; graphic design: Harold Madi (2019)

interviewed over the phone due to scheduling constraints. Interviews took place in January and February 2019.

To contextualize the findings, a description of each of the brand's history follows. Additionally, these in-depth brand descriptions profile a variety of companies that upcycle fashion. This is needed because upcycled fashion is a new phenomenon and the findings from the literature review indicate that the term upcycling has been defined differently by various researchers.

During the research study I analyzed the participants' online presence, including their website, social media, at conferences, numerous global publications, and videos. The findings indicated the participants have been engaged in sharing their brand history and some of their practices in some or all of these forums. Each brand and participant named in the background information section that provides the descriptive statistics section was assigned and identified with the letter A through to letter E for clarity of the findings presented throughout this research study (see Fig. 4).

5.1 *Background Information*

The following section provides descriptive statistics of the participating brands. The type of information provided was adapted from past research studies on the topic of fashion upcycling that also provided overviews of the brands [21, 28]. The brands listed are ordered based as sequenced in the interview schedule. Each of the five participants' brands vary in market, product, what they are known for, and whether

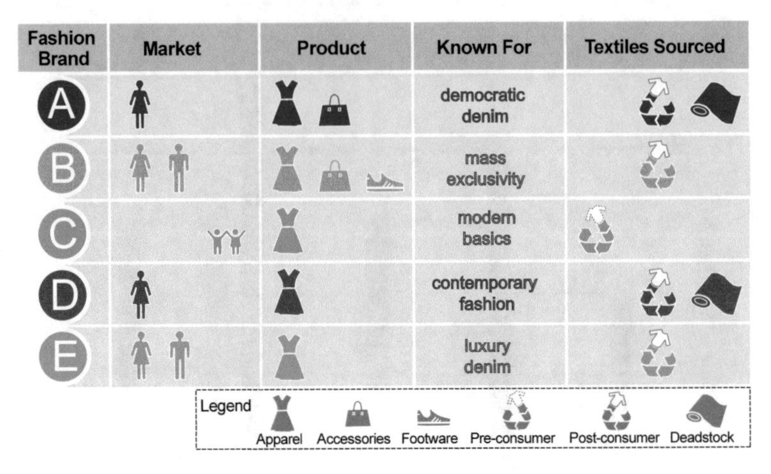

Fig. 5 Brand overview. *Source* Author; graphic design: Harold Madi (2019)

they utilize post-consumer or pre-consumer textiles to create their fashion upcycled collections (see Fig. 5).

5.1.1 Brand A

This independent, Montreal-based denim focused, designer brand, *Kinsu*, sold by over 20 retailers, including the Canadian department store Simon's and the Montreal Museum of Fine Arts. The founder, Ariane Brunet-Juteau, was interviewed for this research study, has been designing apparel for 10 years and has been practising the upcycling of fashionable accessories and apparel for four years. It was while working for a fast-fashion company that the designer became aware of the "power" that they held in the decisions that they made daily. The designer stated that she realized the impacts of her choices, for instance, using cotton versus polyester, a plastic button versus a wood button or faux fur versus real fur for large-scale fast fashion manufacturing could significantly impact the environment. The designer practices fashion upcycling to create women's apparel and accessories using post-consumer textile waste sourced from a local large-scale textile recovery and sorting centre, where they also serve as a board member. At the time of the interview, the designer was planning the launch of a one-day swap event at a private college, in order to expose teenagers to sustainability.

During the interview in the designer's live/work studio, a selection of upcycled denim mittens in shades of blue and black had been arranged on a table (see Fig. 6). Initially, the mittens contrast leather palms and fingertips were made with vintage leather skins from used motorcycle jackets. Regretfully, most wearers of these motorcycle jackets had smoked leaving a scent that was virtually impossible to remove, therefore new leather was sourced for the mittens. The designer also discussed the

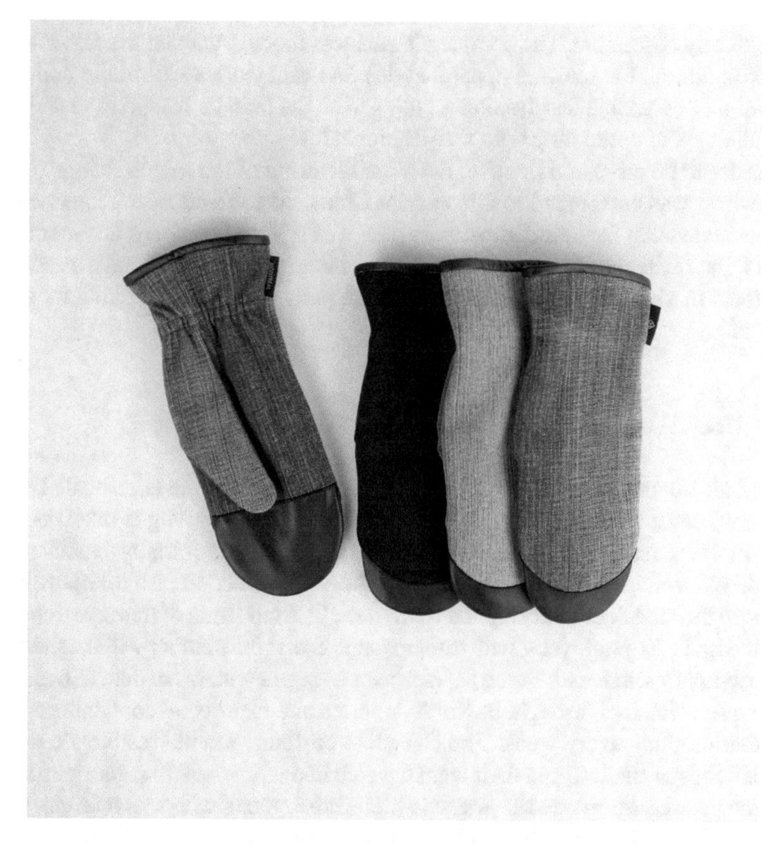

Fig. 6 Brand A's iconic upcycled denim combined with new leather mittens. *Photo* Kinsu Atelier [33]. https://www.etsy.com/ca/shop/KinsuAtelier?ref=seller-platform-mcnav

new, faux fur material for the interior of the mittens and explained that sometimes new materials were combined with those being upcycled to produce functional and saleable products. Decorative surface techniques had been incorporated into past collections. For example, Shibori dyed denim was created as a collaboration with two other local designers. The Shibori technique is an ancient Japanese form of resist dying where the textile is manipulated by clamping, twisting, folding or crumpling prior to the textile being dyed similar to tie-dye [60]. The designer explained why denim was chosen as the main textile for the brand:

> And I have always been a fan of denim, because denim is such a fascinating material. It is the most democratic material in the world, [worn by] women of all ages, men of all ages, it transcends culture, religion, ages, sex. Everybody wears denim. (Brand A)

The designer had been working on solutions in terms of cost associated with upcycling given all of the extra steps involved in the manufacturing process. She indicated that her company is committed to producing fashion products that are economically accessible. The company's business model included the sale of downloadable patterns for selected apparel styles and accessories within the collection. The designer considers fast-fashion retailers to be the brand's competition and noted the challenges of competing with fast-fashion's low-price points.

In order to be more competitive, the manufacturing of the mitten linings (cutting and sewing) was outsourced to a factory in China. At the time of the interview, the designer stated that the brand's customer base was primarily on Etsy, an e-commerce website that hosts independent artisanal and vintage brands. They have also sold at small local markets. Their target customer is an individual who is looking for unique products.

5.1.2 Brand B

This brand, Beyond Retro, launched in 2002, is a division of the large-scale Ottawa-based post-consumer used clothing trading company Bank & Vogue that was established in 1992 (see Fig. 7). The *Beyond Retro* brand was initially launched as a trend-driven vintage clothing retailer. They began to practice fashion upcycling in 2011 with the launch of their Beyond Retro Label.[1] At the time of this research study, the brand had 120 employees, with the expectation that the number will increase. The parent company Bank and Vogue [2] has been engaged with more than 250 charities and private collectors throughout North America and exports 50 containers of post-consumer clothing every week. The Canadian co-founder and creative director for the label, Steven Bethell, and their team have visited more than 30 countries globally to source post-consumer textile-waste [3]. The co-founder, who was interviewed for this study, is considered to be a thought leader and a pioneer on the topic of post-consumer textile-waste and has shared their knowledge at various conferences. The co-founder and two more company employees participated in the Ellen MacArthur Foundation's report titled *A New Textile Economy: Redesigning Fashion's Future 2017* [22].

At the beginning of the interview Bethel described the mission statement for their founding company: "Our mission originally was innovative and relevant solutions to the crisis of stuff." The co-founder discussed the catalyst for creating their fashion upcycled label was identifying a gap in the marketplace, namely their consumers' trend-driven desires that vintage clothing and accessories in their original form could simply not provide as described below:

> And often, our mission in the vintage business is to interpret modern trends as best as we can through use. But at times there are trends that you can't interpret because of the shape, cut, style. You can't interpret through trends, but maybe you can reinterpret or remanufacture a garment so that it can reflect a modern trend and that's why we got into upcycling (Brand B).

The co-founder mentioned they have been referred to as mass exclusivity, which he describes as a category of apparel and accessories where the style is the same, but

[1] See: https://www.beyondretro.com.

Fig. 7 Brand B's Bank & Vogue, an Ottawa based large-scale global wholesaler of post-consumer textiles is the parent company of the Beyond Retro Label. *Photo* Beyond [4]. https://www.beyond retro.com/pages/about-us

the textile may vary in colour, shade and wear. They also discussed the challenges of upcycling for mass versus one-offs:

> The heavy lifting is trying to take a thousand and make a rule of one. Then make one item out of it. But from a retail point of view, it makes an exciting story because if you then now look at my denim snap front skirt, there's fifty of them on a rail with every one of them being different. (Brand B)

The company's production cycle is vertically integrated and includes a trend forecasting team of two and a design team of four based in their head office in the

U.K. This department identifies the direction of the collection, based on research two seasons in advance. Vintage garments are selected during in-house fabric fairs to create unique redesigned apparel and accessories. During the design process, a production manager works alongside the trend and design team to advise them based on the company's analytics and business intelligence, collaborating to combine art and commerce.

In addition to the Ottawa based post-consumer waste sorting company, the brand also owns a large global remanufacturing and production facility in India, allowing for traceability along the supply chain. In addition, the company owns a vintage sorting facility that is located across the road from the manufacturing facility, allowing them to select the items they require for production and return unused garments to avoid waste. The company contributes to the community they are remanufacturing their upcycled collection in by providing aid to the Karuna Girls Orphanage (Bank and Vogue [2]). The women's and men's wear Beyond Retro label are sold via the brand's e-commerce site, Urban Outfitters, Top Shop, and at its nine eponymous shops in the U.K. and Sweden (Beyond Retro [4]).

5.1.3 Brand C

Launched in 2016, Nudnik is a Toronto-based, modern-basics, gender-neutral upcycled children's wear collection that has received accolades from fashion icon Jeanne Beker. The brand has been featured in numerous publications, including *Blog TO*, *Today's Parent*, *The Financial Post*, *Flare*, *The Huffington Post*, *Toronto Life*, and *WGSN* [45]. In 2018, the brand's history as told by the two founders, who are twin sisters, was the feature of a *Globe and Mail* X Volvo series video titled "Exceptional Canadians" (2018). The brand is known for its colourful t-shirts made from pre-consumer textile waste (see Fig. 8). The co-founder Lindsay Lorusso had worked more than 15 years at her father's waste management company, one of the largest in Canada. She mentioned that on a global scale, the category that was most difficult to recycle was textiles, second only to plastics. Lorusso having worked in all facets of the waste management company, discussed her insights regarding materials waste and how that ignited a passion for investigating ways to improve upon problems related to post-consumer textile waste. She discussed how her experience highlighted their concerns about waste:

> So, we spent a collective 20 years working for our father's waste management company. And some of the things that we noticed that were having a really tough time recycling globally was ... plastics, but then secondary was textiles. So, we just started, I mean the idea for Nudnik was conceived about, probably 6 years ago … and we were just starting to really get a lot of the information mainstream about how bad textile waste and pollution was. And from the waste industry we really discovered that there's really not a lot of local or global solutions to handle it. (Brand C)

Initially, the founders created the collection using various types of post-consumer textile waste, mostly adult sized t-shirts and sweatshirts sourced from rag houses. They also worked with local designers utilizing their post-production excess yardage

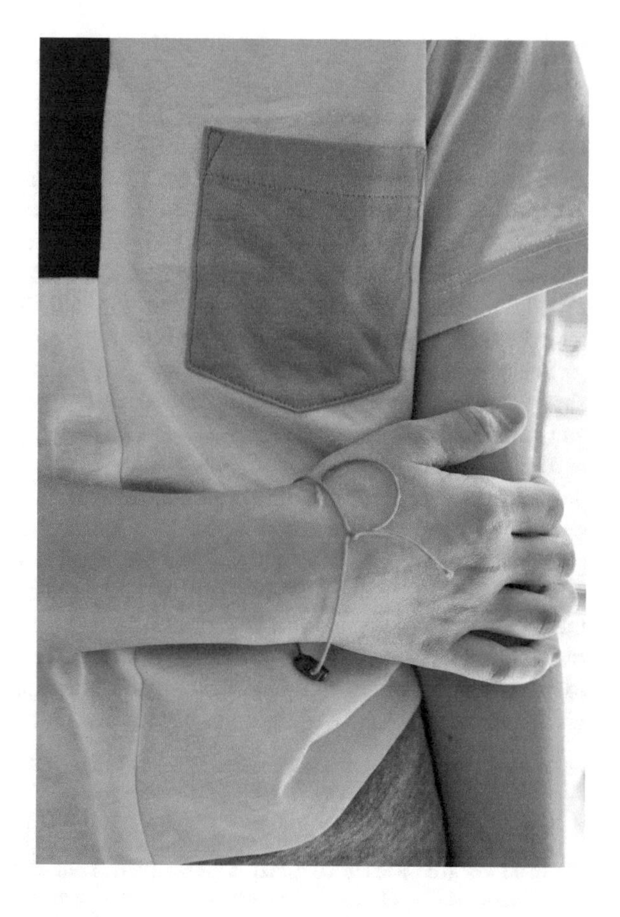

Fig. 8 Brand C's iconic children's unisex "Disruptor" Tee created with pre-consumer organic cotton, threads and trims. *Photo* [45]. https://littlenudniks.com

(end-of-roll textiles) as well as local print shops with misprinted yardage that would be considered textile-waste. Regretfully, none of these sources provided the volume required to scale the number of units produced to achieve an economically viable upcycled fashion business model. The co-founder stated they have since begun sourcing pre-consumer textile-waste that is available in abundance from major apparel manufacturers overseas in Bangladesh, India, Pakistan, Cambodia, and China to overcome the challenges associated with volume. They work with a key coordinator at one of the factories where they manufacture their collection. The collection is primarily available online along with numerous pop-up shops and a few Toronto children's wear boutiques with plans for more trade shows to build the wholesale business.

The co-founder refers to the brand as a think tank that is working toward making a significant impact on the textile-waste problem by being a real example of a scalable upcycled business model. In 2017 and 2018, this brand was accepted into two Canadian business accelerator programs, further validating their business idea of pre-consumer textile waste fashion upcycling.

5.1.4 Brand D

The founder and creative director, Julia Grieve, of the Toronto-based brand Preloved is known as a fashion upcycling pioneer in Canada. The brand was established in 1995 and is known as a contemporary women's collection.[2] This womenswear fashion upcycled brand was a nominee for The H&M Sustainability Award at the 2019 Canadian Arts & Fashion Awards (CAFA). The founder, who was interviewed for this research study, has been a board member and host for *Buy Design*, an event that raised funds for the charitable organization Windfall Clothing Service that provides new post-retail clothing to Canadian families at no cost.

She is a regular expert guest for fashion segments on *Cityline* (a Canadian talk show) and they have their own YouTube channel, The Life, where they share DIY and sustainable projects that repurpose vintage clothing.[3]

Before the launch of the brand, the founder was a full-time model, working for designer brands like Chanel and traveling the world. She discussed the knowledge she gained in the luxury designer space, for instance the importance of fit of a luxury garment, and how that felt to wear. She has applied the same attention to detail, regarding fit, in her collection. She loved wearing vintage clothing and would modify the post-consumer apparel to make it modern and extremely fashionable. Multiple requests to purchase the upcycled clothing the founder was wearing was the catalyst to open the first Toronto boutique with a studio workspace in the back. The founder discussed the importance of creating modern and fashionable upcycled apparel:

> I just always had a passion for vintage clothing, but I felt that the problem was when you wear vintage it can balance into sort of costumesque. If you are not going to modernize it slightly, waistlines are higher or lower or a slit should be slightly different, or hemlines are different. You had to just slightly tweak the vintage clothing to make it modern and that is what I would do, and I loved it, and that is what sort of started the whole idea of it. Obviously taking something old and make it new again is nothing revolutionary, that's the oldest trick in the book, but to be able to make it extremely fashionable and have your line carried in boutiques like Holt Renfrew and Anthropology and all those kinds of places. That's the art behind it I find (Brand D).

In 2014, the brand moved their design studio, retail shop, photography studio, e-commerce, and social media department into the local factory they had been manufacturing in for many years. The brand's team sources vintage clothing made mostly from jersey or sweater knit garments. The founder oversees the creative direction, working with the brand's head designer to develop and edit the collection. The factory owner is an integral part of the design meetings to guide for efficiencies in the manufacturing process of this vertical operation.

Collaborations for collections with large retailers have included Anthropologie, Roots, Indigo, Holt Renfrew, Pink Tartan, The Bay, and Sporting Life. The brand has been featured in numerous international publications and is worn by a number of models, musicians, and actors including Angela Lindvall, Daria Werbowy, Anne

[2] See: https://getpreloved.com/.

[3] See: https://www.youtube.com/channel/UCm0kJUM3HLUzHzTdw-87umQ.

Hathaway, Hilary Duff, Julia Roberts, Kate Hudson, and Kirsten Dunst (Preloved, n.d.). The brand's target customer aligns with the contemporary womenswear market.

5.1.5 Brand E

This luxury fashion upcycled denim company's three co-founders include the brand manager/in house stylist, who is based in Toronto, the creative director/designer, who is based in Los Angeles, and the business director in Vancouver. The creative director and designer, Adam Taubenfligel, who was interviewed for this study, had acquired his fashion knowledge while working in various denim factories located in Italy before launching the Triarchy label with his two co-founders. The founder and creative director discussed their concern for the environment in respect to water consumption and toxic dyes going into the water stream when it comes to denim. They write blogs, create and post videos and speak at conferences on the subject of creating sustainable fashion. The brand received The Fashion Impact Award and, The H&M Sustainability Award in 2018 at the Canadian Arts & Fashion Awards (CAFA). The brand has received accolades from numerous publications including *WWD* (*Women's Wear Daily*), *Denimology*, *Elle Canada*, *Fashion Magazine*, and *The Huffington Post*.

Triarchy was initially launched in 2011, using the standard fashion design model. The denim styles were inspired by original vintage denim garments that were sourced at various suppliers. The designer described one of their sourcing visits that inspired them to use the vintage denim as the fabric for the collection:

> Well, the upcycling came by total fluke. It's quite an arduous process trying to duplicate vintage washes [on denim clothing]. I mean it's something we were doing [in our ready-to-wear collection] so I would spend a lot of time in facilities in LA that have vintage denim [for inspiration]. And so, it was just during one of those trips we were pulling pieces and you know saying how this is all exactly what we want, and then, it was like wait a minute, why don't we just?... And then we're looking at these mountains of denim and just decided what if we just started using this fabric? (Brand E)

The first upcycled denim jacket prototype produced by the company was designed so successfully, that when they went back to the supplier and the brand's manager and in-house stylist was wearing the jacket, they were approached by the owner of a denim factory, who has since become their partner in the company (see Fig. 9).

In 2016, the founders relaunched their company as a sustainable denim brand, moving production to a factory in Mexico City, where 85% of the water used is recycled [56]. The sustainable luxury upcycled fashion collection is under a separate label, *Triarchy/Atelier Denim*. The designer works closely with the production manager at the Los Angeles factory, whom they consider to be an integral part of the selection and manufacturing process. The brand is known for their authentic vintage look and therefore, during the selection process, the designer and the production manager handpick only 100% cotton American made denim to achieve the desired garment aesthetic. The denim is deconstructed, redesigned, and reconstructed and sometimes embellished to create original pieces. Swarovski crystals are the only

Fig. 9 Brand D's unisex upcycled denim fringed jacket. *Photo* Alberto Newton. Model: Lauren Innes [56]. https://triarchy.com/collections/atelier-denim

material the brand uses to add to the perceived value of their clothing. The collection has been exhibited to buyers at the Cabana & Capsule Trade Show in New York City. The brand has been retailed at Holt Renfrew, pop-ups, and on the company's website.

This background portion of the Results section provided a brand overview by reviewing the varied practices that provided multiple experiences from each brand. The successes, challenges, and solutions experienced by the participants will be discussed in the next section, Interview findings.

5.2 *Interview Findings*

The research interviews revealed that only Brand D operated their fashion upcycling business entirely in Canada. The four other Canadian designers who were interviewed engaged at some point in time in fashion upcycling on a global scale with some phases of the design, production, and retailing processes taking place outside of Canada. Interviews with those Canadian fashion designers who operated their businesses globally during the design and production phases were relevant to provide insight into the current state of the fashion upcycling practice. All of the brands sell online and therefore it is likely that they are all selling product globally. This section of the analysis is divided into three main themes—design, production, and retail—that each include multiple subthemes. In each subtheme, successes, challenges, and solutions that were experienced by the participants are discussed.

5.2.1 Design: Fashion First, Sourcing Materials, and Collaborations

Four of the five participants discussed the importance of creating a collection that prioritized fashion first, in that they considered aesthetics to engage consumers. Brand C designs children's clothing and they focus on colourful t-shirts to appeal to a child's eye. They did indicate that they do research trends and that colour palettes are based on trending colours. The words the participants used to describe the type of product they strived to create and the type of product that was successful with their consumers included "fashion, trend-driven, innovative, unique, cool, and surprising." Trend research was deemed critical for the success of the collection, as stated by the participants of the two larger companies, Brand B and Brand D, who worked with a forecasting team that determined key silhouettes, textiles, colours, and print trends.

The co-founder and creative director of Brand B mentioned the word *trend* a total of 30 times during the interview, and they emphasized the importance of trend-driven product and fashion first for the collection. They have created bags sold at retailers such as Top Shop and Urban Outfitters, citing that the consumers respond most to on-trend product. The co-founder and creative director of Brand B discussed the trend and design department and their role within the upcycling design process:

> We have a trend department that chooses the trends two seasons ahead, but chooses the trends from both street style and watching the catwalks. We reflect those trends in our vintage curated pick. But, on top of that, we then identify the trends that we're missing or can't find enough of. Or trends that we feel would be really fun to be able to interpret through the remanufacturing, the upcycling. So, we have a trends department. There's two people that do [source] trends continually. And then there's a team of four that are designers. They come from a true design background, from the London schools. Some have industry experience. Some are straight out of school and one of them is straight from our retail shop floor. And then on top of that, we have a production manager who sits beside them and guides them, against the analytics of, what the business intelligence is saying, in terms of what we need, what is trending in terms of, so there's the arts component, which is the trends. But we also need the business intelligence, [the logistics] of look how many pants do we need, how many shirts do we need, how many skirts do we need. (Brand B)

The co-founder and creative director of Brand B also discussed a recent survey:

> So, it's interesting. We just finished a survey, an internal survey of all of our customers, we ran it for 30 days. We said, what are your buying priorities and obviously fashion and relevance is the first one. And then the second one is sustainability. But I think I would rather make a really cool product. And yeah, by the way, it's sustainable…than try to [only] make a sustainable product. (Brand B)

The co-founder and creative director of Brand C discussed the importance of fashion first and the stigma around sustainable fashion:

> … the look of the product comes first because there is still a stigma around, well, if it's a product that's good for the environment it must look like a paper bag or something like that. (Brand C)

Brand D cuts heart shapes from red textile-waste and appliques them on the outside of the garment or the inside of their garments to surprise the customer. They also applique patchwork hearts created from various shades of red fabric to t-shirts and sweatshirts. This concept evolved as a solution to the problem they encountered of what to do with all of the red textile waste they had put aside in a massive container, since a retail buyer had written on one of their purchase orders that nothing should come in red. The patchwork hearts are now the brand's signature and have expanded as they are now cut from various colours of post-consumer textiles.

The founder and creative director of Brand D discussed the importance of fashion and trends when fashion upcycling:

> I think that's been one of our key elements with our success with Preloved is that we are, I always say we're fashion. I set out to make you look good, saving the planet just happened. Our [business] model is about fashion and then sustainability is the cherry on the top. It's the best part, but because we still stay so focused on fashion and trends and that part it's what's given the brand the longevity. We're not a crafty brand. It's not about taking old stuff and putting it together. ... That's not what we're about. We're about making cutting edge clothing, extremely fashionable, wearable clothing that just happens to have a recyclable element to it. One of the biggest challenges is ... you have to keep evolving. (Brand D)

Brand A's founder and designer discussed the correlation between creating a surprising product and its economic success. They designed mittens with an unexpected material—stretch denim combined with leather—to create a product that surprised buyers and consumers, resulting in one noteworthy sale of a thousand units to a major Canadian fashion retailer. The founder and designer discussed how innovative concepts were fueled by to find solutions that utilized the vast amounts of denim textile waste in original ways:

> The innovation in fashion in the next few years won't be about shape. We're about ethics, style, innovation. And the innovation would be also, aha, we're going to reuse all those clothes that we've produced in the 20th century, in the 21st century. And how in the future we're going to be able to produce clothes that will stay in the loop and we're going to be able to make fashion circular. So that's where the innovation is, so that's [my motivation] for providing my pattern, I'm not really inventing anything. So, people really [relate] to that, and it's a good product. It's good quality. It's really a product that looks like now, and people like denim. So, for them, it's really cool, to see it in some other way that they never seen it [before]. So, for me, my design process, it's really about more like finding a product that will look nice in denim. (Brand A)

Each of the participants stressed the importance of fashion first in creating an upcycled product that will be appealing to the fashion media, retail buyers and the consumers. They indicated that the consumer is most interested in fashionable, trend-driven product and that the sustainable component is increasingly expected by the consumer to be part of the DNA of the brand.

5.2.2 Design: Sourcing Materials

All five of the designers cited challenges with garment sourcing that included finding large volumes of materials that are consistent in fibre, in the size of the piece of textile within each garment, in print or pattern, wear, colour, and shade. The largest sizes available provided the most viable textile use of fabric for production. Each designer emphasized that the design phase began at the sourcing stage. Designer participants elaborated on sourcing locations, these included: post-consumer textile waste from sorting centres and rag houses and pre-consumer textile waste from standard fashion design companies or fashion apparel factories to acquire post-production off-cuts or end-of-roll textile waste. Each of the brands had evolved in terms of the type of post-consumer or pre-consumer textile waste they sourced based on the volume of materials available, and that influenced the aesthetics and values of their brand's identity.

Brand A designed using post-consumer indigo and black stretch denim and non-stretch denim products only, as these are abundant in supply. All were sourced locally from the largest textile recovery and sorting centre in Quebec, along with individual donations directly to the designer. Denim textiles align with the brand's philosophy of democratic fashion for all genders, ages, and cultures. The designer considers the pattern shapes and sizes they are designing in terms of what could be cut from the leg of a pair of jeans. To save the selected denim that has the brand's desired worn aesthetic for production, the designer has implemented a strategy to utilize other companies' deadstock end-of-roll textiles to create the multiple sample prototypes required to create the desired fit and style. The designer of Brand A discussed their sourcing process:

> People just want to get rid of their clothes, and this is where the need is. So, people are always very happy to give their clothes … and people give me jeans. And, I also go to Certex, which is one of the biggest clothes sorting centre in Quebec. And they're amazing. They sort 7,000 tons of clothes every year. They have all those bins with jeans, so I feel like the design process starts here. But I need to have an idea of what I want to do. With the mittens, you need jeans with some stretch, so it's more comfortable. With the bags, I work with 100% cotton. So, it depends, so when I go there, I start sorting with 100% cotton and the heaviness and lighter weight jeans, jeans that stretch and always black and indigo. And then I always think of my product in a way that the pattern [piece] will fit within a leg of jeans. So that's another point, because when you have fabrics, there's no problem here, but working with upcycling, you always have to think your product in a way that it's going to fit into one leg of jeans. (Brand A)

Brand B has its own sorting facility across the road from their factory in India, a country that has large volumes of post-consumer textile waste, providing the brand with a variety of materials in volume for their trend-driven collection. The co-founder and creative director discussed their sourcing process:

> So, what the process of it goes through is, let's say the trends department will appoint the trends and fabrics. So, we'll say, let's just take a couple of easy ones, you know, and these won't be relevant, but I'll pick them anyways. We'll say, look tartan is going to be big this fall. And, let's say denim, you know a certain hue of denim is going to be really big or white denim is going to be really big come spring. So, the trends department, just like a typical house in the U.K. would do, instead of going to the show in Paris and going to the fabric fairs, we have our own internal fabric fairs. So, what we do is we'll say, okay, we know that these are the five key fabrics for A/W 19. What we'll do is we'll then start pulling all of that fabric and we'll build volumes of it. So, we'll pick 10,000 pieces of plaid. And then, literally the team will then talk to the trend department and we'll say okay, what shapes are

> we looking for, what things are we looking at designing, if we were short on skirts or short on pants. And then based on that, they'll take the fabric fair, the commission of fabric, and then we'll start trying to mine out certain shapes or products out of that fabric. (Brand B)

Additional challenges were discussed around the selection of garments with "good" damage versus "bad" damage based on the company's aesthetic, this judgment call required specialized training for those selecting the garments that serve as the raw material. The co-founder and creative director of Brand B discussed the challenges of training employees to recognize the company's aesthetics of damage to maintain a consistent look:

> On the picking side, teaching an aesthetic of what is good damage and bad damage can be challenging. … The western aesthetic is something that would be beat up and kind of wear on it's even better. How do you translate what is valuable distressed versus shit distressed? (Brand B)

In 2018, Brand C changed its business model from utilizing post-consumer textile waste to utilizing pre-consumer textile waste. Before this, the designers were only able to produce capsule collections in Toronto due to the limited supply of textiles available from local designers. The co-founder discussed the challenges with sourcing materials in volume:

> The biggest challenge we had with upcycling or using textile waste in general to make new products is that it's really hard to source a consistent flow of material. And when you don't have that you kind of get stuck in this place where you might be doing these capsule collections based on just some of the smaller volume of material that you might have sourced. But it's hard to say that you can keep that production going, and so, we found that very challenging in terms of even being able to sell the product. We wanted to be able to supply some boutiques and things like that because they showed a lot of interest, but we were never able to guarantee that we would be able to keep giving them enough of what they really liked. So, I think that is one of the biggest challenges in upcycling is that you're trying to always source a consistent flow of this waste material to be able to output a product that's consistently made for people that expect and want to buy it when they want to buy it. (Brand C)

The co-founder of Brand C discussed their solution to this challenge in that it was to shift their sourcing to apparel factories overseas in Bangladesh, India, Pakistan, Cambodia, and China, where the amount of pre-consumer waste is abundant, allowing them to create a scalable upcycled product stating, "one factory could output about 150 tons of this specific cutting waste bi-weekly. So, there's really no shortage of it." When the factories are cutting standard collections, a marker is made that places all of the pattern pieces together, as close as possible to maximize fabric utilization, however, there are gaps in the layout in between the pattern pieces and the size of this textile-waste is well suited to making children's apparel. The company chose to work directly with factories who were producing organic cotton fashion apparel to source pre-consumer sustainable textiles. The co-founder discussed the built-in advantages to using pre-consumer textile waste from the apparel industry, the fibre content is known, colour palettes have been researched and are trend-driven, textiles are already sorted by colour and the laundering phase to clean the textiles is not necessary. The brand benefits from the prior research and development conducted by those designing the originating collection. They then create their own colour palettes from the various pre-consumer waste. The brand also uses end-of-roll threads from the factories to match the textiles sourced. Since they are working with the factory to source the textile, they also work with factories to manufacturer the product. The brand's garments are designed without notions such as zippers or buttons to create a completely upcycled collection.

Initially, Brand D worked with various types of vintage garment-textile waste selected from rag houses. The brand was not allowed to select from the vintage section, as that was saved for the boutiques who had already built a relationship with the supplier. The designer described their first visit to the supplier, with their team when they were directed to select from the bright coloured undesirable prom dresses in the corner. This limited access was the catalyst to the upcycled collection as the dresses were able to source was all they had to work with. The brand decided to redesign the prom dresses into tank tops and skirts. They have since built relationships with other suppliers and can select the vintage garments that are harmonious with their brand's aesthetic and values. The brand primarily sources French terry, jersey, and wool sweater knit garments to create their upcycled collection. The founder and creative director discussed the importance of consistency when sourcing garments for the upcycled collection:

> A wool sweater is a consistent garment. Our goal when you work with upcycling and you are wanting to mass sell this, the key word is consistent. You need to create a garment that has consistency. A buyer does not like the word "assorted." A customer buying a garment online wants to receive close to what [the sample style] they ordered. If you are going to upcycle and you're not looking at mass, that's when you can have a lot of creativity. ... That's not my business model. My business model is mass. So, you're constantly looking for consistency. Things that have consistency. So, a wool sweater has consistency. (Brand D)

Brand E is known for an authentic vintage aesthetic, and they select from multiple suppliers in LA to specifically 100% cotton American made denim (see Fig. 10). The creative director and designer of Brand E discussed the challenges their company faced when sourcing:

> In LA you have so many vintage suppliers it's insane; warehouses upon warehouses of stuff. We have mountains to go through. I used to source whenever we had an order, and then it really became difficult, we were running into this problem, where we would not really have the fabric in stock and then we would get an order and it would be difficult to get the fabric. So, I started buying in bulk. So, the storage locker in our building in downtown LA, is a literal pile of vintage stuff. It's obscene, but at least now whenever we need to make something we just go downstairs. (Brand E)

The challenges discussed by the participants during the design process in the materials sourcing section included having immediate access when needed to large volumes of consistent raw materials. Depending on the type of material required for the season and style, it may need to be consistent in fibre, in the size of the piece of textile within each garment, in print pattern, worn aesthetic, colour, and shade. The participants have implemented various systems to solve these challenges including vertical integration that includes a sorting centre for materials sourcing, building relationships with the owners of rag houses so that they are provided with

Fig. 10 Brand D's 100% cotton indigo denim with fringe trim. *Photo* Alberto Newton. Model: Lauren Innes [56]. https://triarchy.com/collections/atelier-denim

a premium selection of vintage or sourcing pre-consumer textile waste at fashion apparel manufacturers.

5.2.3 Design: Collaboration

Three types of collaboration were discussed during the participant interviews, including in-house collection collaborations, artisan alliances, and partnering with other brands. These collaborations provided some solutions to the many challenges presented when creating an upcycled fashion collection. The in-house collaborations consisted of trend and design teams working with the production and pattern making teams simultaneously. This is different from the standard fashion design model where the trend and design team would work in a different silo and pass off the design only once it's completed to the pattern making and production teams. The artisan alliances included partnerships with other makers such as local Shibori dyers. The brand partnerships discussed were with either major brands or major retailers.

Two of the participants had adopted a vertically integrated business model and discussed the importance of the collection's collaboration within the brand, including all of its stakeholders, the trend forecasters, designers, production managers, and business managers (see Fig. 11). Vertical integration is a business model when multiple phases of production are completed by the same company versus some phases being contracted out to other companies [50]. Through working together, the stakeholders were able to create fashionable products while working in the most efficient way possible by anticipating the challenges of working with post-consumer textile waste and developing solutions as a team during the design phase. In the Design: Fashion First section discussing Brand B within the topic of trends and fashion, the co-founder and creative director also discussed how the various stakeholders within the company collaborated to run it more efficiently. The founder and creative director of Brand D discussed the benefits of moving their company to within their manufacturer of many years and how they work as a team (Fig. 12):

> We moved into our manufacturing plant. We have worked with the same manufacturer for probably about 15 years. We work with many, but our main one is WS & Co or Redwood, they have two names. And it's the owner of the factory. ... He is just wonderful. He has been a big part of our manufacturing process over the years. ... He is involved in almost all of our production meetings…amazing talented man, he's a part of our design meetings because how we design will help how we manufacture, which will help how we cost our product. (Brand D)

Brand A's designer had engaged in artisan collaborations in order to transform the denim material in new innovative ways through the sharing of techniques amongst crafters globally. The designer discussed a few collaborations with specialist artisans:

> I did a collaboration… with a Shibori specialist. ... She uses whole natural indigo, and she went to Japan multiple times to learn about the technique. So, it is really the ways to fold and the many times you dip into the indigo… I am really interested in natural indigo. ... And I also worked with another girl that is an embroiderer. So, we did a couple of things and she's making my labels. She had the industrial embroidery machine. ... I also contacted people on Instagram. A girl in Africa, she is making necklaces with t-shirts, that I would like to do the same technique with denim. So, this is another way that I'd like to bring Kinsu, is really collaborating with crafters all across the world, recycling denim. ... I would really like to

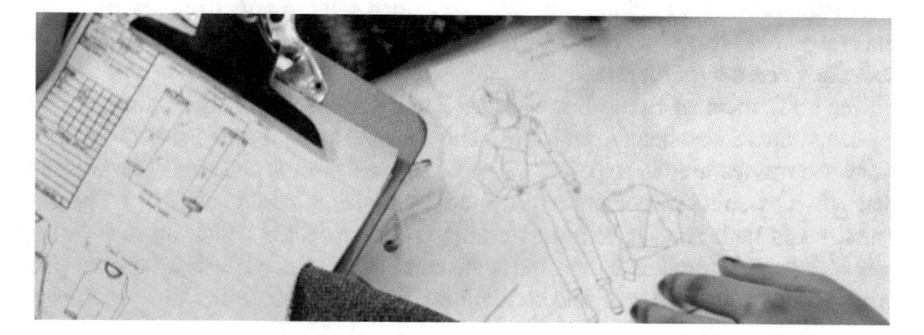

Fig. 11 Brand C moved their company into one of their manufacturer's for vertical integration and in-house collaboration. *Photo* Preloved (n.d.). https://getpreloved.com/

Fig. 12 Brand C moved their company into one of their manufacturer's for vertical integration resulting in increased efficiencies. *Photo* Preloved (n.d.). https://getpreloved.com/

collaborate and, present the material in some other way and different embellishment. (Brand A)

The Shibori specialist had grown indigo plants locally to create the dye for the brand's iconic upcycled denim mittens thus aligning with the brand's sustainability ethos (see Fig. 13). The artisan alliances provided solutions to the challenges of the upcycled denim always appearing the same, by elevating the textile-waste and providing financial gain for those involved.

Additionally, two of the participants described brand partnerships as being an integral part of the business model whether they are made public or as a private label collection created for major retailers.

The co-founder of Brand B collaborated with large-scale brands elevating their participation in sustainable practices. The brand's most recent partnership with Converse was featured in *Hypebeast*, a popular men's fashion and streetwear website that provides editorial content and has an e-commerce component [7].[4] The co-founder and their team have been working on the upcycled denim project for the past 3 years. This project demonstrates the concept of mass exclusivity as each pair is made of various shades of selected denim. The partnership with Converse Renew

[4] See: https://hypebeast.com/2019/8/converse-renew-denim-chuck-70-high-low-ox-beyond-retro-release-information.

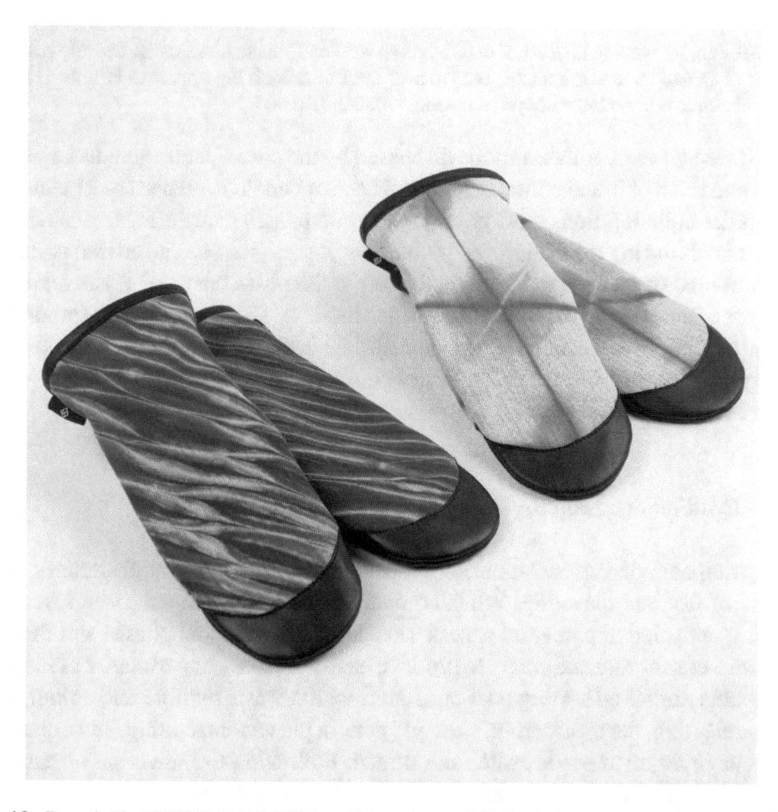

Fig. 13 Brand A's KINSU X INFUSE collaboration of Shibori-dyed upcycled denim mittens. *Photo* [33]. https://www.etsy.com/ca/shop/KinsuAtelier?ref=seller-platform-mcnav

Denim's upcycled iconic Chuck Taylor styles, were released on August 22, 2019 [53].

Brand B has also been successful in creating private label bags and backpack collections for major brands sold at Top Shop and Urban Outfitters. The founder and creative director of Brand D discussed collaborations with major retailers and working with their design teams to create capsule collections, including Roots, Indigo, Holt Renfrew, Sporting Life, and Anthropologie:

> So, we would work with these brands and work with their design team and then because they would reach out to us because they loved the concept and we want to create a product that was right for their customers, but still had a bit of Preloved in it. It was so exciting. I still love it. That's our big thing. We launched this season with a big collaboration with Sporting Life.... [It is] all about the great lakes of Canada, Lake Simcoe, Lake Rosseau, and got little recycled hearts and the line is deadstock material. It's super cute. I love collaborating with other retailers. That's probably my favourite part because you get to understand what their customer likes. The way they work. (Brand D)

The co-founder of Brand C discussed the possibility of engaging in Canadian made collaborations in the future:

> And for us, we want to do a lot of collaborative work in Canada like smaller capsule collections of Canadian made products, but we also want to make a big impact to be able to be a true example of a scalable upcycled business model. (Brand C)

The three types of collaborations discussed by the participants included, in-house, artisan and retail or brand partnerships have been of benefit to all parties. The in-house stakeholder collaborations have provided the companies many efficiencies resulting in lower costs during the design and manufacturing processes. The artisan collaborations provided ways for the designers to elevate the post-consumer waste textiles by dyeing or embroidering to create new surface interest. The partnership collaborations with other brands and retailers provided further insight into a better understanding of the retailers' customer. The collaborations provided increased visibility for all parties involved.

5.2.4 Production: Labour, Cost, Scale and Textile Quality

The participants discussed labour, cost, and scale as the biggest challenges in the practice of fashion upcycling. When compared to typical garment assembly, manufacturing upcycled apparel and accessories demand additional phases and thus it is more time-consuming and costly to produce each accessory or garment. For instance, participants discussed sorting post-consumer textile waste by fibre and colour, laundering, selecting, deconstructing, strategic pattern placement, cutting, and organizing by colour in order to shade match the thread. Following the selection of garments from the supplier during the design process, the second phase of selection occurs during manufacturing. Garments with the desired amount of wear or in the case of denim, with "whiskered" marks, are selected and the pattern pieces for the style strategically placed, with the worn or "whiskered" area in the location communicated by the designer on the sample style. The designer co-founder, creative director, and designer of Brand E discussed the challenges with the cost of manufacturing their upcycled collection:

> We do it all in downtown LA, because that is where the expertise is and that is where we are. But truthfully, it is just abhorrently expensive. Like when I sell anything from the Atelier line, it is just covering our cost, and that is the truth. (Brand E)

Each garment requires deconstruction before cutting, adding to the production time. All of the brands treat the disassembled garments as the fabric that they cut from to create the redesigned garment (see Fig. 14). One of the participants stated that the practice of zero-waste upcycling might produce more creative results, but it is far more time-consuming and generally only allows for the production of one-of-kind or couture pieces. Zero-waste is the practice of creating a garment wasting none of the textile typical in standard fashion design and the technique may also be applied to upcycled fashion design [51]. Zero-waste techniques aim to reduce textile waste through various approaches such as the jigsaw method, or the tessellation method that employ non-traditional geometric pattern shapes and often multiple garments

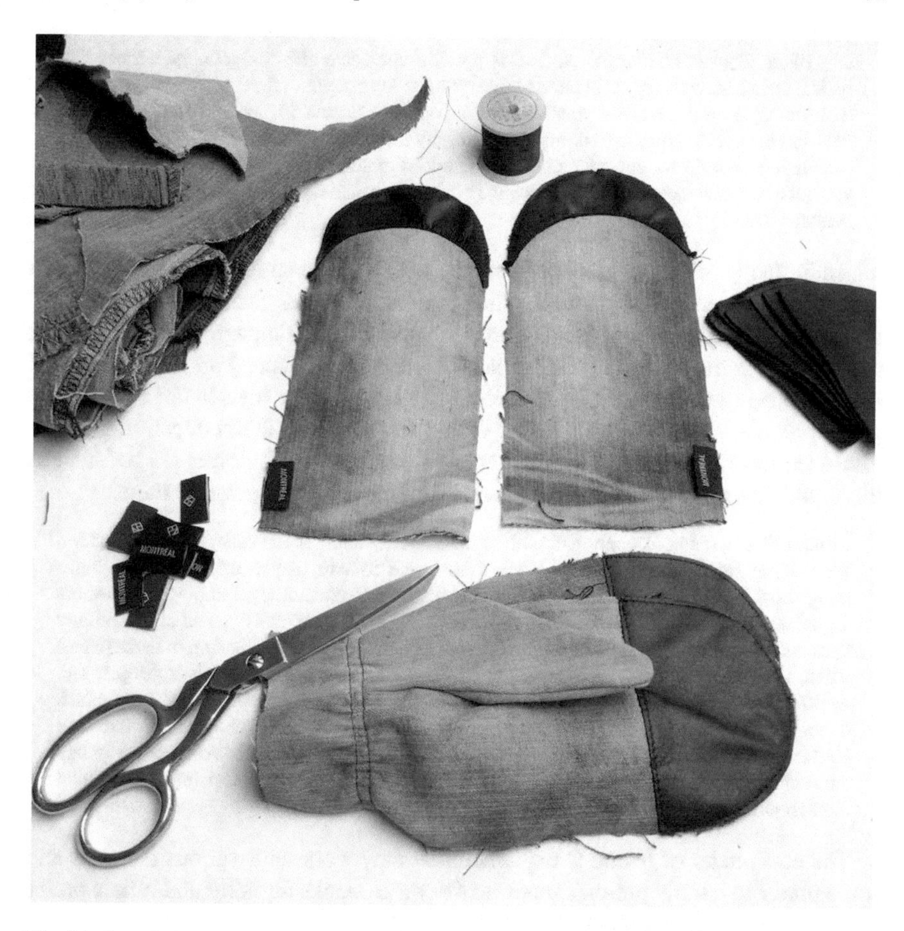

Fig. 14 Brand A's iconic upcycled mittens during deconstruction and reconstruction. *Photo* [33]. https://www.etsy.com/ca/shop/KinsuAtelier?ref=seller-platform-mcnav

within a marker layout leaving no gaps of fabric. Standard fashion design processes waste can range between 15 and 30% of the textile in a traditional pattern marker.

The founder and creative director of Brand D also discussed cost as being a challenge when fashion upcycling versus standard fashion design with the additional steps required: "Another deterrent is cost. Cost for sure. It is a labour of love… if you buy a bolt of fabric, there is no deconstruction. There is already from the-get-go, one extra step" (Brand D). Finding efficiencies within the manufacturing process was cited as the biggest challenge by all of the participants, as most apparel producers are used to cutting off the roll. The challenges in cutting from the fabric available in sourced garments include size and shape limitations. This requires additional labour related to decision-making within these parameters.

One participant discussed how the cutting challenges affect the scalability in fashion upcycling:

> One of the biggest challenges we had is how do you scale and compete against new? A typical cut and sew factory will have a table, you can lay out your fabric and be 50 deep. And then you lay your pattern, use your knife cutter, and you can cut 50 pieces all at once. Where with the upcycling you are continually cutting by the ones. One of the biggest hurdles we have had is how do we cut efficiently and on-speed to compete against new. Because once you get to the sew factor, you bundle your 12 components, and you can hand it to the sewing person. (Brand B)

Additionally, placing the pattern piece for each style strategically to utilize the worn or whiskered areas in the upcycling of denim requires each cutter to be trained to recognize aesthetically "good" damage versus "bad" damage from the brand's perspective. Some of the brands use hand laser cutters to speed up the process, but no one is using any other cutting technology because the investment exceeds the profitability. The co-founder of Brand C who manufactures their collection overseas discussed technology to aid in the processes, but said the factories are resistant as this could mean job losses where apparel manufacturing is a primary industry:

> I found that the idea of using technology here to cut, some of the manufacturers that we were doing some of the work with are more eager to start maybe using some of those things now. But I find overseas they are really resistant to technology. I know they do some digital cutting. But what I found in conversations is that they are really scared of technology because technology can basically mean the replacement of their people. And in these places where apparel manufacturing is a major primary source of income for their people, and workforce for their people. But they are really hesitant to take on anything that will replace them. Because there really are not any other jobs. So, there has been some really interesting conversations had with some of the people that we are working with. And actually, they are coming to ask us the same questions. Like, would you go over to using an artificial intelligence (AI), a robotting cutting system? Or robotting manufacturing system. (Brand C)

The co-founder of Brand C explained that they work with various factories and some prefer to cut the pre-consumer textile waste stacks digitally and others prefer to cut the pieces individually.

Organizing by colour to match thread was cited as a challenge since it is also time-consuming. Brand A developed a system that utilized old beer bottle boxes with 24 separate compartments to organize each pair of the collection's denim mittens by colour and shade during production not only for thread match but also to ensure the same amount of "wear" was on the right hand and the left hand. Brand C's solution for matching the thread was to cut the same colour trim for a single style.

All of the participants discussed the additional labour required during the fashion upcycling processes and the associated increased costs compared to standard fashion manufacturing processes. The larger brands with vertically integrated companies discussed the benefits associated to labour and costs within the design and manufacturing phases. Some of the brands have developed techniques to aid in the time-consuming process of organizing the textiles by colour for the sewing processes. All agree that there is room for improvement in the efficiencies of labour and costs.

5.2.5 Production: Textile Quality

The participants discussed a number of challenges associated in working with post-consumer waste including stains, odours, holes, and stretched out areas. The participants have implemented a variety of techniques to solve these challenges. The designer of Brand A discussed the challenges of working with the vintage leather that they were sourcing and why they decided to work with new leather skins in combination with the upcycled denim for their iconic mittens:

> For me, when you are upcycling, it is like you are adding a level of risk with the quality, because you don't have standard material, and actually you have to work to standardize a material that is not standardized. And this is with the sorting. Okay, this is this weight, it stretches, it has a hole, I cut beside it ... or I put it in the middle depending on what I do if I want it, if it fits with what I'm making. But with mittens you don't want holes, because you want them to be warm. And for me that job of standardizing the material that I was upcycling it was enough with the denim.
>
> And I felt like the leather, it was harder, because also with the denim, it's always pants. I can count okay I can make four pair of mittens average in a pair of jeans ... I need to have two hundred pairs of jeans to do my production. Same thing with the bags, ... I can do one bag with one pair of jeans. So, I can count like this, but with the leather, it's some skirts and jackets and there's all those seams. And often its people who smoke that wear leather.
>
> So the leather would smell ... sometimes the skin is dry ... and also, there is so much waste, all the knees, the elbows, it will stretch out with the leather. So, I could not use this. All the zippers, the lining, the interlining, so I was finishing with small pieces of leather and a mountain of garbage, of stuff that I could not use. (Brand A)

Given the constraints of holes, stretched out areas, the smell of cigarette smoke, the various types of garments, and the multiple seams typically found in vintage leather garments, the designer no longer used vintage leather but new leather skins instead, combining new leather and post-consumer denim within some of the products for their collection. The participants have implemented a variety of laundering processes for the purposes of cleaning the post-consumer materials. Stains and odours in garments sourced were mentioned by a few of the participants as problematic. One of the designer participants discussed the challenge of stains, their solution was to launder all of the vintage product that they had selected from the supplier before sending it into production. Initially, they sent all of the vintage clothing out for laundering, but the eventual vertical integration of this designer's company included a laundering facility onsite reducing the time it previously took to send the vintage to an external laundering facility. The laundering would allow the team to determine if stain removal was successful, and if they remained, to remove the garment from production. Additionally, the laundering process proved that the integrity of the quality of the garment was in excellent condition and if the textile did not hold up, the garment would be removed from production.

Some of the brands had discussed agreements with rag houses to provide them with bales of used clothing that were already laundered. One of the participants discussed that their laundering takes place in a patented laundering process that uses 90% less water than standard laundering. Another participant cleans all of the garments at a laundromat, and they plan to implement a more sustainable laundering practice using ozone washing machines, thereby reducing water consumption.

To ensure the post-consumer textiles are of high quality, each brand had implemented rigorous hand selection processes and various types of laundering practices including on-site laundry, pre-laundered bundles provided by the rag houses, and patented laundering processes that use significantly less water than regular washing machines.

All of the designers called out scale as the number one challenge within the production phase. During the interviews, they discussed small volume production as not being economically sustainable. The inherent challenge was how to mass-produce and to make tens of thousands of an upcycled style using available sources of textile waste. The leading solution discussed was to design using garments with material consistency, such as denim, that is available in abundance at the sourcing level. The participants indicated that to achieve scale they must find even more efficient solutions to the additional phases of labour required in the fashion upcycling model that include sorting the post-consumer textile waste by fibre and by colour, selecting, deconstructing, strategic pattern placement, cutting, and organizing for manufacturing. The participants working within a vertically integrated company and location indicated that this had aided in improving the scalability of the company but acknowledged more efficient techniques need to be developed to increase volume.

5.3 *Retail: Acceptance, Price and Brand Narrative*

The participants discussed the challenges of acceptance by the fashion media, retail buyers, and consumers within the realm of upcycled fashion. The participant from Brand D discussed the improvement they have seen in the acceptance of upcycled fashion during the past 25 years but suggested that another decade of exposure may be required for upcycled apparel and accessories to be accepted by the mainstream stating, "Honestly, I still think after 23 years, it is still a bit of 'oh it is used.' … It has gotten a lot better, but I still think that is one … at the consumer level … there is a stigma. …. We need another decade of brands like Reformation."

The co-founder, creative director and designer of Brand E discussed the challenges when selling to retail buyers regarding consistency in upcycled collection: "It is problematic because there's no, well, there's consistency in fit, not in the colouration … but for the most part, if people are buying it for the story, it really does not matter." The co-founder and creative director of Brand C discussed the positive press they have received and the specific challenges of children's wear in the fashion media:

> The press has been extremely receptive and again children's wear is not something that is put in fashion magazines, or really talked a lot about. There is this stigma that kids wear is just an extension of adult wear and that it does not hold weight on its own. But what people really focused on in the press is that of how we do what we do. And you know we have been in *Today's Parent* and most of the major Canadian news outlets. So, what excites us now

is being able to break into the U.S. market and see if we can do something similar there. (Brand C)

Before 2014, the fashion media provided scant coverage on sustainability and next to nothing on fashion upcycling. Since 2014 fashion upcycling has been provided more visibility via multiple fashion publications with the participants in this study. The participants in this study have been featured in fashion publications such as *Denimology*, *Elle Canada*, *Fashion*, *Flare*, *The Globe and Mail*, *The Huffington Post*, and *WGSN*.

Some major Canadian and international retail buyers have embraced unique upcycled fashion products designed by the participants interviewed for this research study. The challenge one of the participants cited with some of the retail buyers is that the sample they see at the buying appointment is the product they expect to receive in-store, in terms of fit, the textile fibre, colour, and pattern. While some buyers love when there is variation in the assortment, others do not. Another participant indicated that it is a challenge to get the buyers attention to view the collection so that they may grow their brand. All of the brands sell directly to the consumer via e-commerce on their website therefore bypassing any retail buyer resistance.

The majority of the participants had recounted stories of customers being excited about the upcycled fashion product. All of the participants reported that they had not received any negative consumer responses directly about the utilization of textile-waste for fashion apparel and accessories, but there is some hesitation when it comes to price as discussed in the next section.

5.3.1 Retail: Price

In an earlier section, labour and associated costs were cited as one of the main challenges during the production phase by all of the participants. The costs are ultimately reflected in the retail price. Some of the brands that have vertically integrated supply chains discussed how that helps to decrease the costs and those savings are then passed on to the consumer. Two of the five participants cited the challenge with some customers' expectations that the price of fashion upcycled accessories and apparel should be inexpensive. To mediate this assumption, the founder and creative director of Brand D emphasized that the textile is vintage, but the design is new in an attempt to address the consumers' perception regarding price:

It is just using vintage materials to make that. That was a huge issue that I had seen 15 years ago was really difficult to get people to understand. Is it vintage, is it used? No, it is not used. That garment is brand new. You could buy that garment in extra small to an extra-large. …The fabric is what is second hand, but the garment itself … that's what is the big difference with Preloved. We also work with deadstock materials and new fabrics and blend the vintage. Sort of a new part of what we do. You need to create a great product that's sustainable. That's how you become a success today. (Brand D)

The co-founder, creative director, and designer of Brand E discussed their approach to price perception in the luxury market: "You need the optics of vintage to justify the price. We usually just work with Swarovski, because the Atelier line already has such a high price point, we have to add excessive embellishment to kind of show the buyer."

Three of the participants discussed the importance of accessible consumer pricing to provide democratic fashion and to be able to make an impact by redesigning the large volume of textile waste, thereby diverting it from landfill. The co-founder and creative director of Brand B discussed the price that is acceptable for their consumer. They discussed the challenges of producing the product locally at accessible price points; therefore, some of the product is manufactured overseas:

> For upcycling to be truly successful, or for us to make an impact or move the needle at all on the crisis of stuff, our average price point has to be no more than 24 pounds. Call it 35 Canadian dollars. So, if it is going to be accessible, then we have to figure out, okay, how do we execute within the margins we typically want so that we can land an item on our shop floor for $35? So, the reason why this is important is we start with our target price. This is what we need to achieve. Product has to hit the floor at this price. (Brand B)

In an earlier section, the designer of Brand A mentioned that they had begun to sell their patterns online. This was a way to provide access to what might be considered an expensive product by some consumers. Brand A's solution was to adopt a disruptive open-source fashion business model by offering for sale, the same items from their collection as DIY (Do It Yourself) downloadable patterns, priced at a fraction of the price, allowing the items to be more accessible to the consumer (see Fig. 15). Niinimaki and Hassi [44] describe open-source fashion as a business model that designers use to sell their patterns and the sewing instructions to the consumer. This open-source sharing portion of the business model also provides passive income to the company:

> I'm also selling my patterns online with tutorials, and if you want to do it yourself, go get it, then do it. And wow, you're selling your pattern. It is such an opposite way. In fashion we are so, my God, this is my pattern. But we are not thinking this way … I feel we are sharing now. In fact, Kinsu is becoming ready-to-make focus instead of ready-to-wear. I have also in the past month created two licenses for the sewing patterns, a personal one and a commercial one, as I was beginning to have demand from crafters who wanted to sell the goods, they make with Kinsu patterns. (Brand A)

All of the participants discussed that the time-consuming practice of fashion upcycling with the extra required phases during the design and production processes often leads to higher prices at retail. Each brand has been working toward finding efficiencies at various phases of the fashion upcycling process from sourcing through to the retail.

5.3.2 Retail: Brand Narrative

All of the participants indicated that their companies are promoting the sustainable practices they engage in either on their website, on their garment labels, or on their hang tags. The designer of Brand A reports that the customer is looking for transparency: "More and more people are looking for more sustainable or looking for

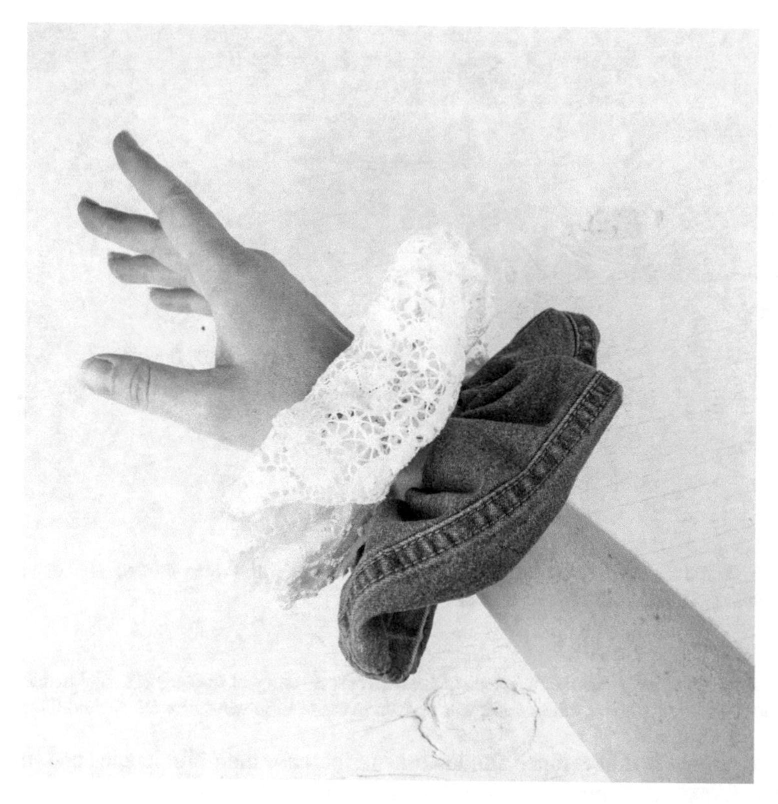

Fig. 15 Brand A's DIY patterns are available as a Personal License or a Commercial License. https://www.etsy.com/ca/shop/KinsuAtelier?ref=seller-platform-mcnav

having confidence in the person that is selling the product. More people want to know where it was made, by whom. These are the questions that the customer right now is more interested in that, the transparency."

Brand A's designer discussed that while enrolled in a business program, they were told not to talk about sustainability because people did not want to know about it. The participant believes that not talking about sustainability is an outdated way of thinking.

Brand C's labels are screen printed on to the garment with their tag line "Negative Waste. Positive Impact" communicating the sustainability narrative to the consumer visually and eliminating the extra textile that would normally be used for a label. The co-founder and creative director of Brand C discussed how their labels are made (see Fig. 16):

> But, for right now, for this next collection we're putting out were using just these end of roll threads, trims, waste, scraps and then waste inks for the little screen prints label inside and then the tag. And again, we're using a screen-printed label inside the garment as well, just inside the t-shirts as opposed to a cotton one. And again, it's just the intention that we don't

Fig. 16 Brand C's iconic children's unisex "Disruptor" Tee with screen-printed label using waste ink. *Photo* [45]. https://littlenudniks.com

> need to necessarily use another piece of fabric when we can just screen print right inside the garment. So, it's about minimizing waste, even in those little ways as well. (Brand C)

Each brand had developed frameworks to increase their efficiencies and through the participation of this study have shared their knowledge that may further advance the practice of up-cycling. All of the participants communicate the various sustainable practices their company engages in via blogs, video, television, social media, and by speaking at industry or academic conferences to further the brand narrative.

This section's findings of the qualitative data included challenges and solutions that were organized by theme of design, production and retail. A summary chart was created for each of the three main themes: design, production, and retail. Each chart provides an overview of key findings in relation to challenges and solutions experienced by the participants during the upcycling processes (see Figs. 17, 18, and 19).

This section presented the results of the qualitative research study that formed the basis of this chapter. All of the participants identified that for the product to be desired by the consumer, most importantly, it must be fashionable. All of the designers indicated that there must be consistency in the materials used for each style, and the materials must be available in large volumes to scale the business. All of the companies were involved in various types of collaborations. Within the manufacturing theme, all of the participants cited labour and costs as a challenge when manufacturing upcycled fashion. All of the companies had implemented cleaning processes at the beginning of the manufacturing phase. Increasing the number of units per style was cited as one of the biggest challenges encountered by the participants during the production phase to scale the business. Each founder had different

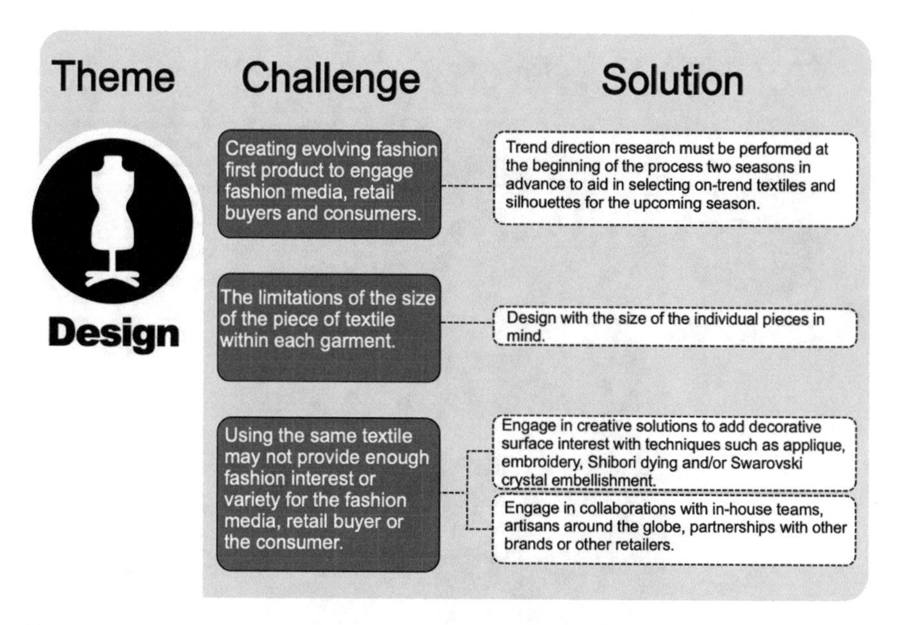

Fig. 17 Design: Summary of findings on the challenges and solutions in upcycling. *Source* Author; graphic design: Harold Madi (2019)

experiences at the retail phase within the category of bricks and mortar, but all of the brands sell their collections via e-commerce.

6 Discussion

This section discusses the results using the same sections of design, production, and retail. The findings from this research study revealed three main subthemes within the Design theme, including the importance of fashion within the product, challenges associated with sourcing materials, their volume and consistency, and collaborations with other brands or retailers. The three main challenges within the production theme were labour, cost, and scale; and textile quality. The retail theme included the challenge of acceptance associated with the fashion media, retail buyers, and consumers; price, and brand narrative. The findings also revealed that some of the design, production, and retail is undertaken in other countries on a global scale. The findings suggest three types of upcycled processes, including: trend-driven, utilizing post-consumer textiles; textile-driven, utilizing post-consumer textiles; and textile-driven, utilizing pre-consumer textiles. Summary charts were created for each of these three processes and a standard fashion process chart adapted from Han et al.'s [28] study (see Fig. 20). These more detailed charts provide insight into the various phases of upcycled fashion (see Figs. 21, 22, and 23).

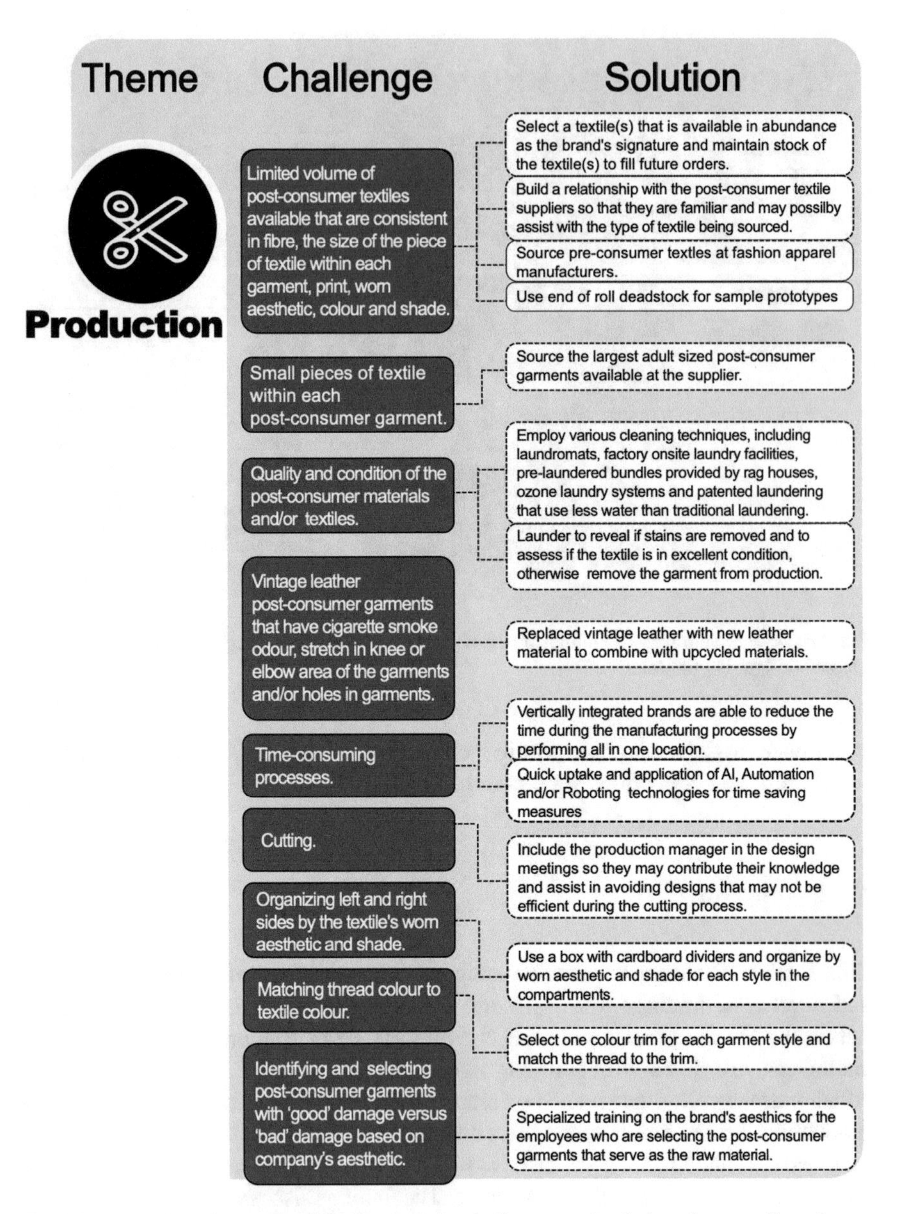

Fig. 18 Production: Summary of findings on the challenges and solutions in upcycling. *Source* Author; graphic design: Harold Madi (2019)

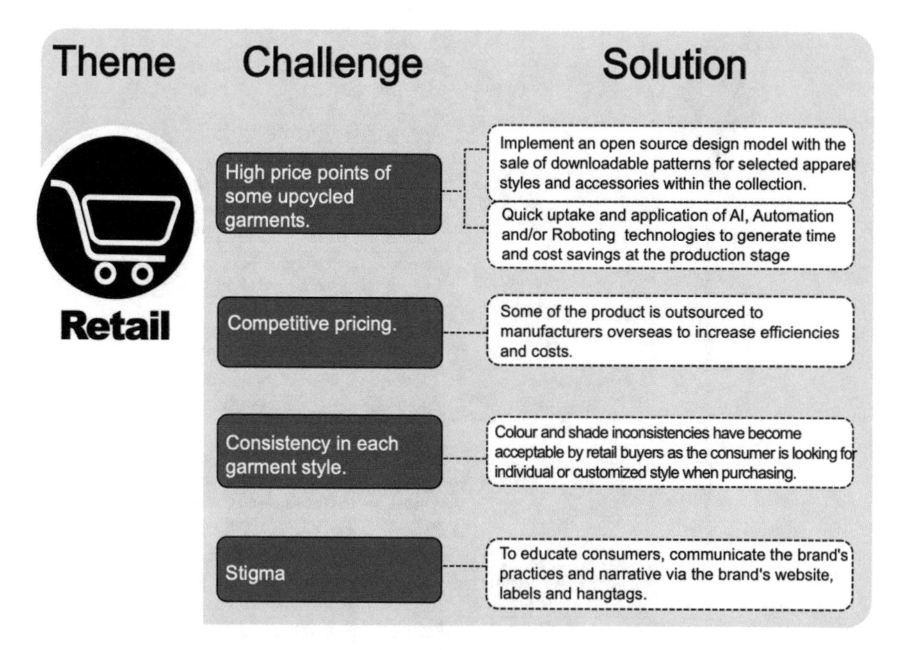

Fig. 19 Retail: Summary of findings on the challenges and solutions in upcycling. *Source* Author; graphic design: Harold Madi (2019)

An additional chart was created to compare the differences and similarities of each of the three fashion upcycling processes described by the participants versus the standard fashion process within the themes of design, production, and retail. The comparative chart is an abbreviated version of all of the standard and upcycled processes (see Fig. 24).

The pre-consumer textile driven upcycled fashion process is similar to the standard fashion process in the number of phases. The post-consumer textile processes, both trend-driven and textile-driven, include the extra phases when working with end-of-consumer garments.

6.1 Design: Fashion First

Fashion first is an integral part of the design brief for the three types of upcycled fashion processes. Few retail buyers and consumers will purchase fashion products only because they are sustainable, thus, retail success is dependent on a sustainable product assortment that is fashionable. Similar results were found in Streit and Davies [54] study, where the participants indicated the product had to be design driven. However, in Han et al.'s [28] study, research in relation to trends, colour, fabric, and style and consumer preferences were the second phase following post-consumer textile selection at the beginning of the design process. As seen in this

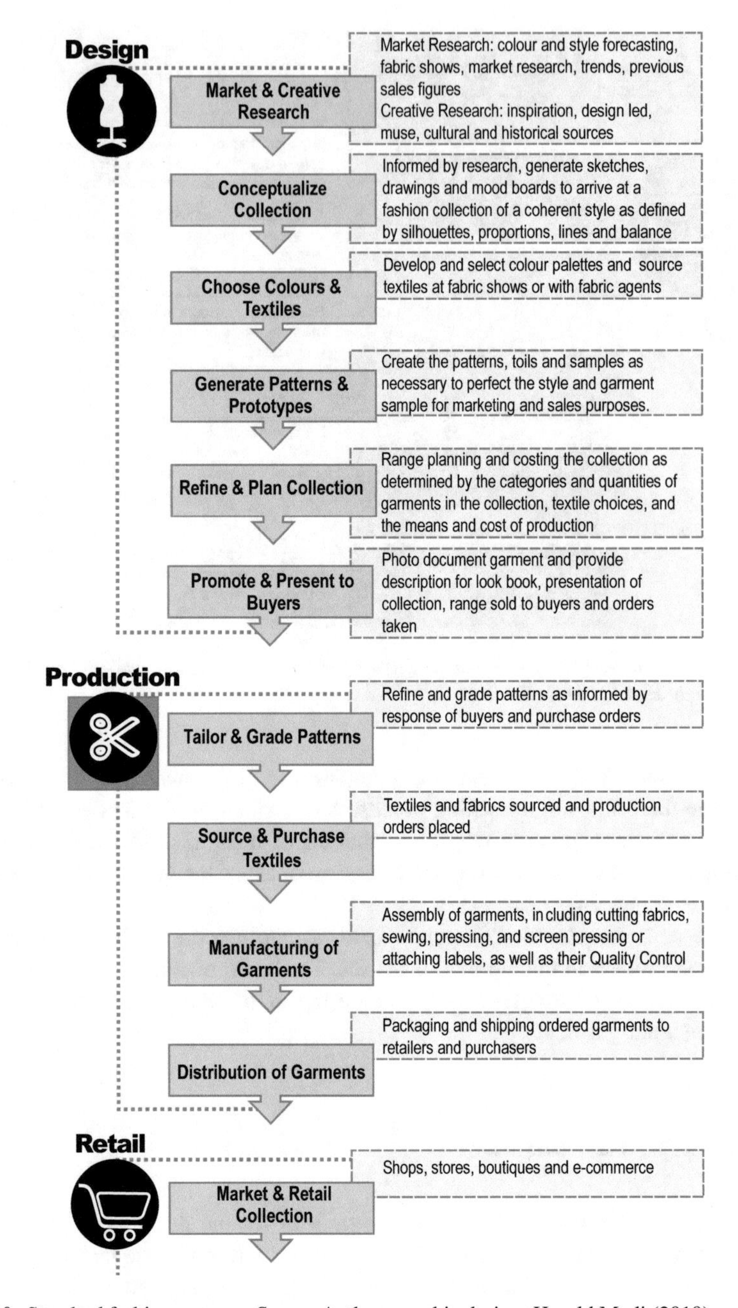

Fig. 20 Standard fashion process. *Source* Author; graphic design: Harold Madi (2019)

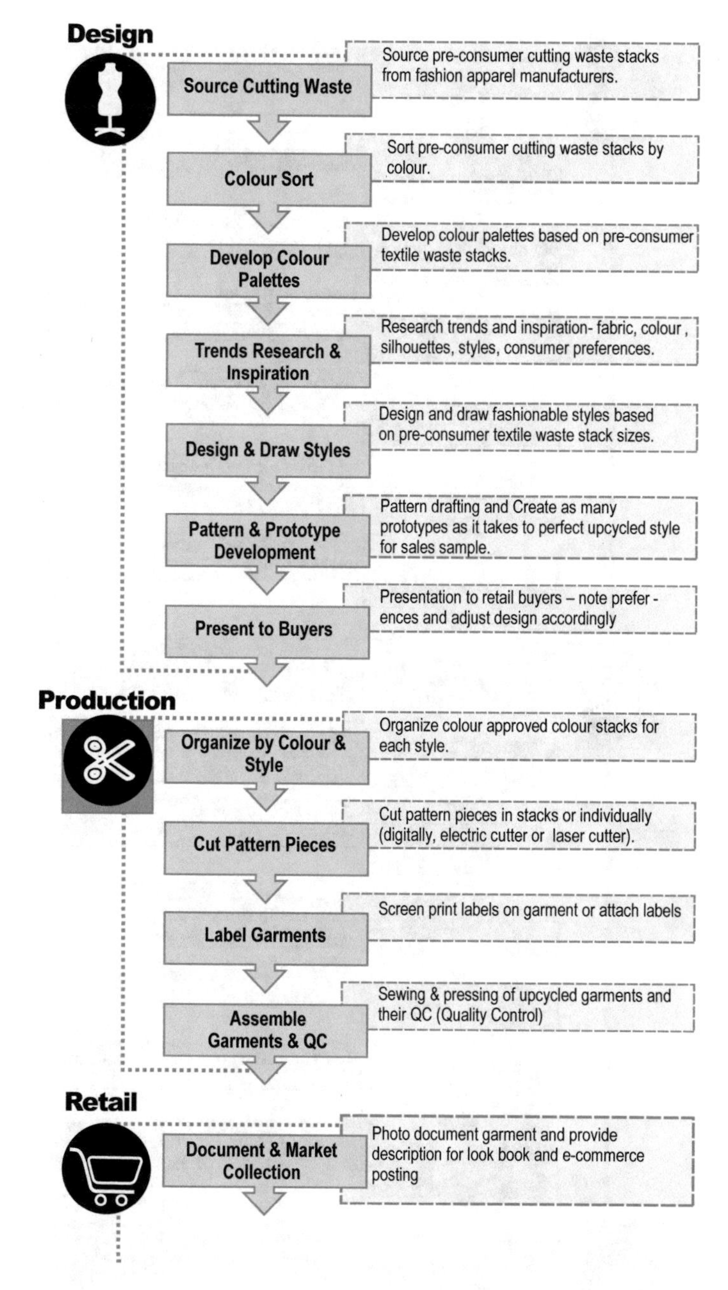

Fig. 21 Upcycled fashion process—Textile driven using pre-consumer waste. *Source* Author; graphic design: Harold Madi (2019)

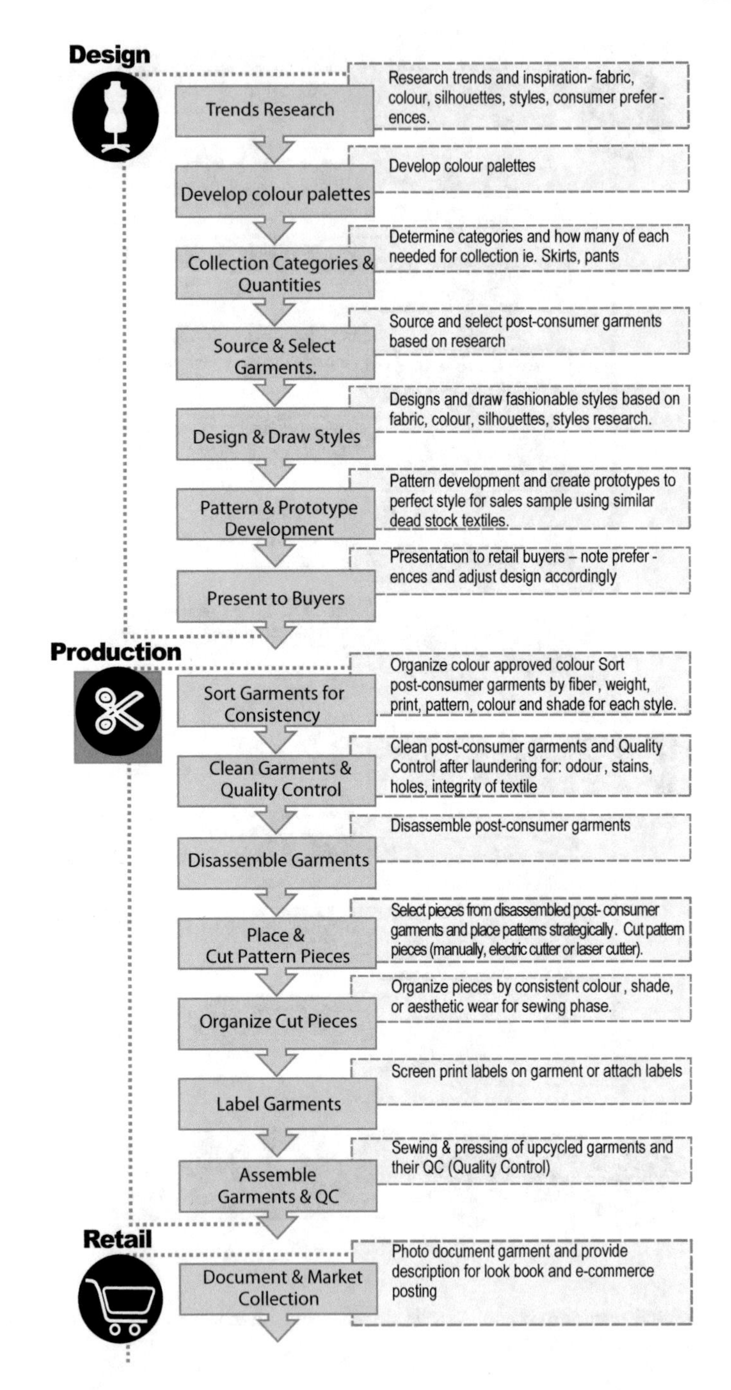

Fig. 22 Upcycled trend driven post-consumer waste process. *Source* Author; graphic design: Harold Madi (2019)

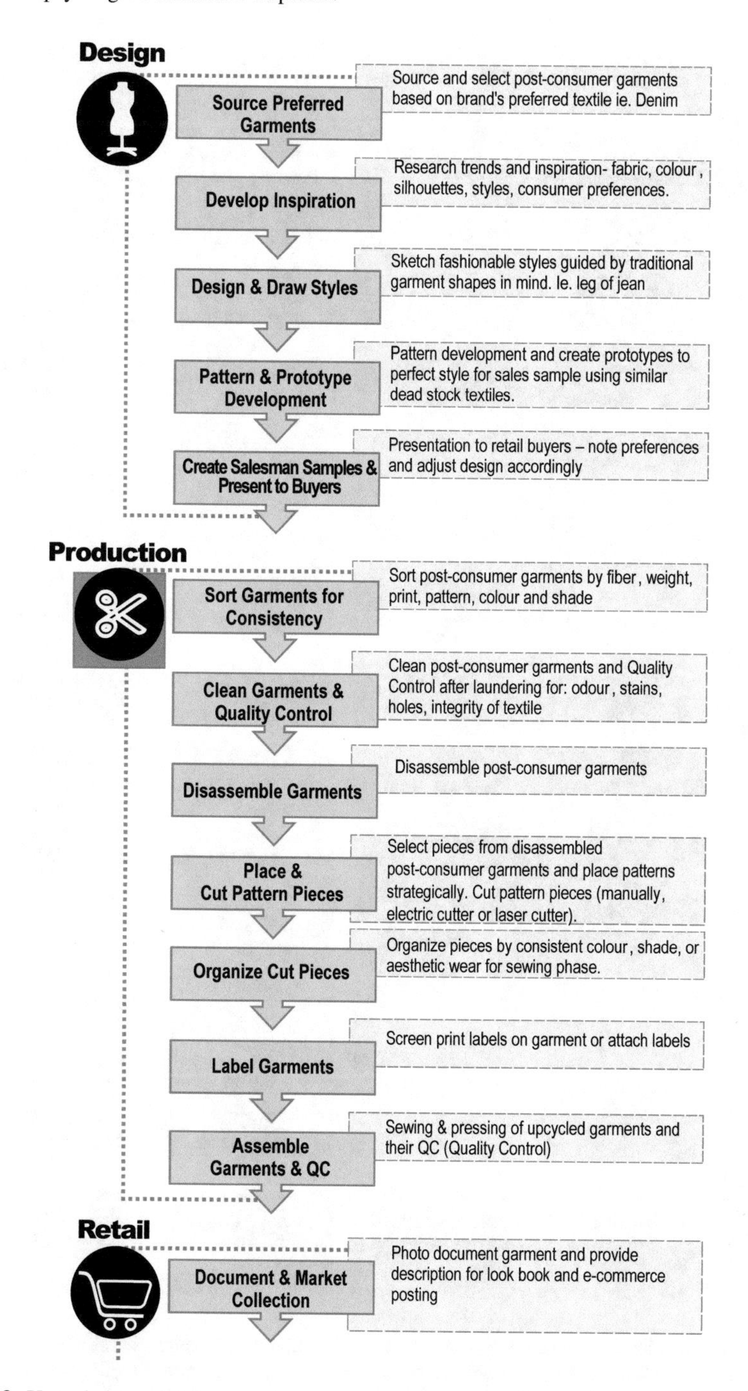

Fig. 23 Upcycled trend driven post-consumer waste process. *Source* Author; graphic design: Harold Madi (2019)

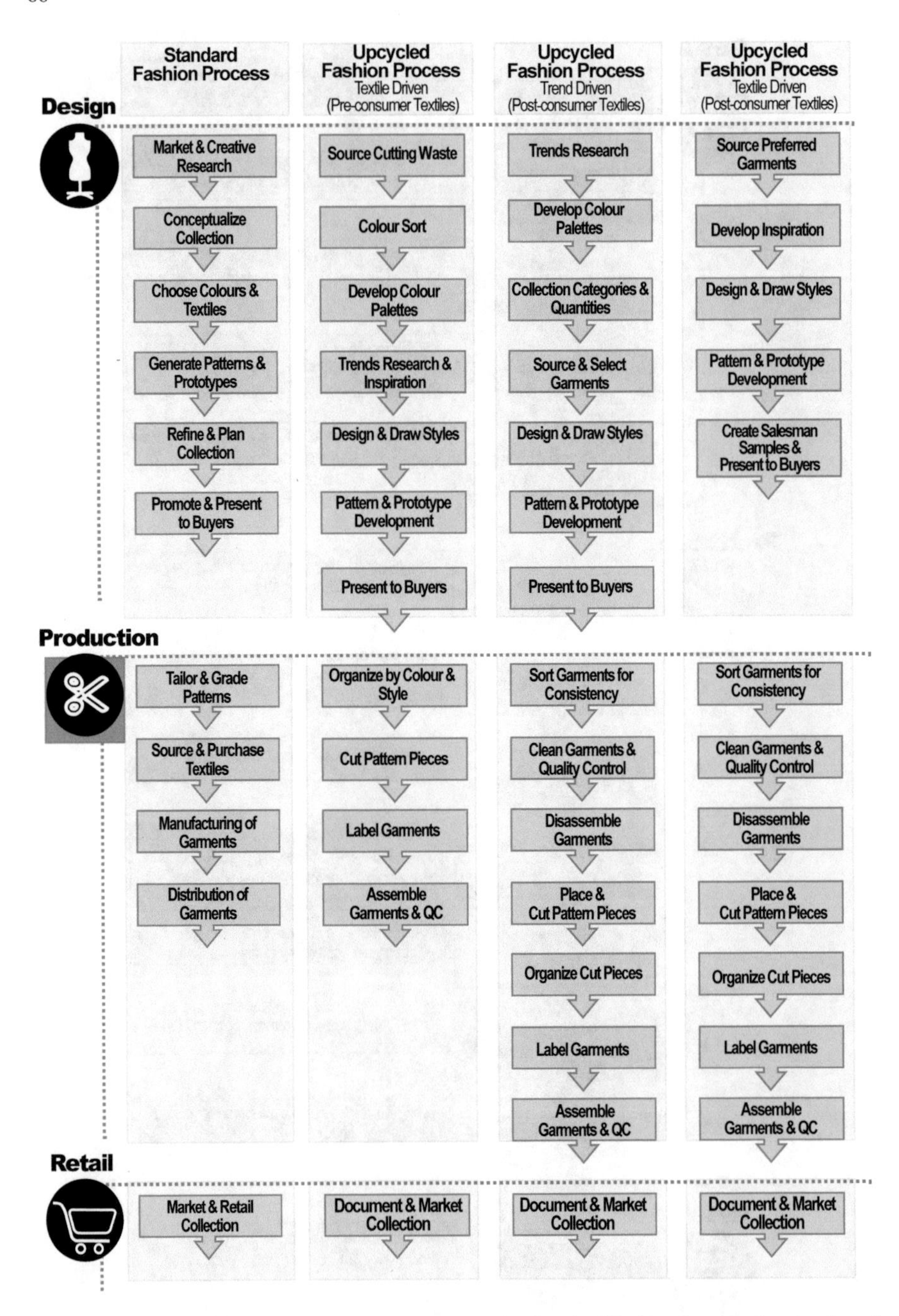

Fig. 24 Design: Standard versus upcycled processes. The Standard Fashion Process chart is adapted from Han et al. [28]. The three upcycled processes are the results from this study. *Source* Author; graphic design: Harold Madi (2019)

study and in Han et al.'s study, there are multiple process orders depending on the brand and the brand's manufacturer.

6.2 Design: Materials, Volume, Consistency

Acquiring consistency in the volume of material derived from textile waste is a universal challenge for designers creating upcycled collections as supply is limited to availability of stock at the various suppliers. This finding is consistent with past studies (e.g., [11, 21, 27, 28] and the companies that participated in this study also indicated the same challenges with materials, volume, and consistency. It is important to note that fashion designers who upcycle denim are the exception. They do not encounter challenges with volume as denim is available in abundance given that people of all ages and genders wear and dispose of garments made of this material.

Although the base textile is available in significant volume there are still challenges associated with consistency (fibre, colour, and pattern). In response, designers have created and implemented various systems, such as utilizing vertical integration in their business model by sourcing from their company-owned sorting plant. Similar to findings in Han et al.'s [28] study, some buyers are accepting of variation in upcycled collections as this provides mass customization, a concept that is increasing in popularity with retailers and consumers.

To solve the issue of acquiring volume in consistency, one of the companies sourced pre-consumer textile waste directly from fashion apparel manufacturers overseas. Creating a collection with a focus on one specific but widely available textile such as jersey or wool sweater knit was another strategic approach.

6.3 Design: Collaboration

Fashion upcycling companies are engaged in collaborations with artisans and retailers to create an innovative product. The brand collaborations are not only a means to increase the economic viability and the visibility of the brand, but also to raise consumer awareness regarding fashion upcycling. Those that strategically align their collaborations with other sustainably driven artisans or partner with retailers working toward increasing sustainability within their company are generally successful. This was highlighted in the findings of the study done by Dissanayake and Sinha [21], where they suggest collaboration with other designers, retailers and commercial waste companies. This is to advance innovation where companies in the study have had success through stakeholder collaborations. Han et al. [28] also advocate that fashion upcycling designers collaborate with large-scale retailers.

6.4 Manufacturing: Labour and Cost

Fashion upcycling is a fairly new process within the fashion design space, and therefore, efficiencies within the system to integrate the additional phases required in the production process have not been fully developed and this impacts labour costs. Labour and the resulting costs have been highlighted as one of the biggest challenges in fashion upcycling [11, 21, 47]. Although garment disassembly and cutting are possibly the most challenging and labour-intensive phases in manufacturing, incorporating affordable technology may increase economic efficiencies.

6.5 Production: Cleaning

Cleaning is an important additional phase in the upcycling process, and one that has several solutions to minimize the impact on labour and costs. Participants in this study recommended many solutions. The various cleaning processes discussed to ease cleaning post-consumer textile waste included: on-site laundering facilities, agreements with rag houses for pre-washed bundles, ozone washing machines, and facilities that use 90% less water for laundering.

6.6 Production: Scale

The biggest challenge in fashion upcycling is in achieving economically viable mass production to create a scalable business. More efficient solutions for disassembly and cutting must be developed. Vertical integration of fashion upcycling companies is vital, efficiencies in the production process will lead to scale given that many of the phases may be conducted simultaneously. This requires collaboration among all of the stakeholders (trend forecasters, designers, production managers, selection teams, pattern drafters, drapers, cutters. and sewers). Cassidy and Han's (2013) study indicated that it was mostly niche fashion companies with relatively limited production volume that performed upcycling. Past researchers [21, 28] performed interviews with upcycling brands that included companies that were noted as "micro" to high-profile brands who were engaged in collaborations with large retailers. Direct comparison with the companies in past research studies versus the companies in this research study may not be made as each study published different types of information in relation to market category, type of product, what the brand is known for, the type of textiles sourced, collaborations, where production was based and with whom the interview was conducted, I would propose that the Brand Overview chart (Fig. 5) created for this study be expanded on and utilized for future research studies so that the area of fashion upcycling may be analyzed further. The findings of this study indicate that there has been some growth in fashion upcycling to include companies that mass produce.

6.7 Retail: Fashion Media, Retail Buyers, and Consumer Acceptance

Contrary to past research [11, 54] the participants in this study indicate that fashion media, buyer and consumer mindset have shifted toward more acceptance of upcycled fashion. Past academic literature, educators, and some in the fashion space indicated that discussing sustainability may deem the brand unfashionable. Increased discourse on the topic of sustainability, the circular economy and fashion upcycling will raise consumer awareness. Streit and Davies [54] exploratory study "Sustainability Isn't Sexy" indicated that all seven of the designers whom they defined as ethical discussed challenges associated with promoting their brand. Cassidy and Han [11] suggested that media could be of assistance in shifting consumer mindset in the area of fashion upcycling. The results from this study suggest that there has been an improvement in acceptance from the fashion media, fashion buyers and consumers during the past 25 years but there is still more to be done.

6.8 Retail: Price

Although all of the five participants practise fashion upcycling, each company has an individual brand aesthetic that lives in different market categories and is directed at a range of target customers. In general, customers have price expectations for upcycled fashion apparel that is comparable to the prices they are used to paying for standard ready to wear fashion. For example, the mass/fast fashion consumer expects to pay certain prices regardless of the product being upcycled or not. This is also true for contemporary luxury designer collections. One notable exception regarding price expectations is that some consumers believe that upcycled products should cost less because fabrics are not new and cost much less than new textiles.

Offering the consumer various ways to access the designed product is worth considering. Niinimaki and Hassi [44] discuss an open-source design and co-creation approach where the designer sells their patterns and directions online for the consumer to construct themselves. The findings from this study are similar, the designer from Brand A sold their DIY patterns with construction instructions online so that the consumer might create their own upcycled fashion. The designer of Brand A recently had multiple online requests from crafters asking if they may use the patterns for their retail business, leading to the creation of commercial use patterns. Niinimaki and Hassi [44] also advocate actively engaging the consumer, as this promotes emotional durability (see also [12] to upcycled fashion items, leading the consumer to use it for a more extended period. Further rationalization of price may be addressed via brand narrative, as discussed in the next section.

6.9 Retail: Brand Narrative

A brand's narrative in the fashion space is essential, however, there are differing opinions about whether or not to discuss sustainability. The fashion landscape has changed dramatically over the past few years, and although previous discourse advocated not to discuss sustainability, there has been an increase in the importance of purpose-driven products for the Millennial consumer [29]. Given that fashion upcycling brands are reusing post-consumer or pre-consumer textile waste thereby diverting it from landfills, it is essential that the brand communicate this to the consumer. This elevates the perceived value of upcycled apparel and may drive business for the brand increasing the likelihood of economic success. The results from this study indicated some designers had been advised not to mention sustainability and these findings were similar to past research [54]. Binotto and Payne [5] argued that communicating the upcycled component as a narrative for the brand can aid in the fashion collection's perceived value. The Paris-based Maison Martin Margiela Artisinal collection is an excellent example of brand narrative meeting perceived value [40]. Past research findings have indicated that communicating the narrative about the history and lived experience of the upcycled textiles can also increase perceived value [5, 9, 25, 32].

A key recommendations chart has been created based on the analysis and discussion of the results of this study (see Fig. 25). These guiding strategies may further the practice and encourage more designers to uptake fashion upcycling.

7 Conclusion and Recommendations for Further Research

This concluding section includes contributions, implications, and directions for further research. The purpose of the conclusion is to elaborate on the themes of design, manufacturing, and retail within the practice of fashion upcycling and how they might advance within the professional and academic practice within fashion through the suggested further research.

7.1 Conclusion

This study contributes new insights to the current state of the upcycling design practice within fashion, and it highlights strategies for designing upcycled fashion

Fig. 25 Key recommendations for fashion upcycling. *Source* Author; graphic design: Harold Madi (2019)

apparel and accessories. Based on the literature review, few researchers have interviewed the founder and designer of brands who are practicing fashion upcycling, and the majority of those studies were conducted in the U.K. [21, 27, 28, 54]. Brand C has broken new ground in its approach to source directly from the fashion manufacturing industry in order to acquire volume in their preferred textile of 100% organic cotton. This technique is the least time consuming of the three upcycled processes noted in this research study, contributing a new upcycled process to be used in future research. Diverting excessive textile waste in landfills that produce emissions and cause harm to the environment can be mediated by advancing the practice of upcycling. Exposing designers to the viability of upcycling through education will encourage more implementation of the practice. Upcycled collections are a viable part of a circular economy for fashion apparel and accessories, thereby reducing the environmental impacts when compared to standard fashion manufacturing processes. This is important knowledge for educators, students of fashion design, fashion designers, and for future academic research.

7.2 Recommendations for Further Research

This study investigated the challenges and successes of creating upcycled fashion apparel or accessories from the designer's perspective. The qualitative findings indicate that the most significant production challenges were disassembly and cutting as they are labour intensive and thus costly. To remain competitive and to scale the business, labour costs must be reduced. I would propose that the comparison chart created in the Production: Scale section, might be utilized for future research studies so that so that the same type of data may be collected.

Future research that focuses on upcycling and the use of technology would provide needed production strategies to ease labour costs, for example the implementation of existing Artificial Intelligence (AI) and robot manufacturing technology to improve efficiencies during the cutting process. Further investigation into the cutting phase of upcycling is required, specifically if there are any affordable technologies that exist. If technologies are successful and implemented, then the implications for the fashion space would be those entire workforces would need to be retrained, shifting from manual operations to AI or robotics. In recent years, a number of postsecondary fashion design programs in Ontario, Canada have included fashion upcycling within the curriculum. Future research could investigate fashion design curriculums at post-secondary institutions on a global scale in regards not only to fashion upcycling, but also the implementation of sustainable practices, open-source design and co-creation models. Further implementation within the fashion design programs will allow for more practice and exploration by students that may contribute to the development of more fashion upcycling strategies. Workshops led by fashion upcycling leaders could be developed and offered at public and private institutions to train fashion design professionals who are in the industry. Another recommendation for future exploratory research would be to include empirical studies to investigate new design principles for end-of-consumer textile products redesign and how these principles might be introduced in fashion design curriculums and to fashion industry professionals.

Upcycling is an emerging practice that is gaining more interest from fashion designers, the fashion media, retail buyers and the consumer. The designers who have been creating upcycled apparel and accessories for more than 15 years have proven the viability of this sustainable design practice. This chapter investigated the challenges, solutions and successes and encountered by founders, creative directors and designers of fashion upcycled apparel and accessories brands. Each participant discussed the systems they have implemented during their time practicing upcycling to increase their brand's efficiencies and through the participation of this study have shared their knowledge that may further advance the practice of up-cycling. The challenges and solutions charts developed provide best practices for apparel and accessory designers who engage in fashion upcycling, for educational purposes to engage future fashion designers, and for the purpose of future academic research.

In conclusion, this research study was undertaken due to the climate crisis and increasing amounts of textile waste being sent to landfills, specifically the waste that is contributed via the fashion space. Since I began this research study in 2018, the numbers of town and city councils that have declared a climate emergency have increased, indicating that we, the citizens of the world must continue to find solutions to the issues that are contributing to the crisis. The central aim of this research study was to determine the challenges, solutions and successes of upcycling to advance the practice, thereby diverting textile waste from landfill and reducing harm to the environment.

References

1. Anguelov N (2016) The dirty side of the garment industry: Fast fashion and its negative impact on environment and society. CRC Press. https://doi.org/10.1201/b18902
2. Bank and Vogue (2019) Our story. https://www.bankvogue.com/our-story/
3. Bethell S (2019) Steven Bethell, President and Partner–Bank and Vogue, Beyond Retro Label: About. LinkedIn. https://www.linkedin.com/in/steven-bethell-64341b73
4. Beyond Retro (2019) Our vintage. https://www.beyondretro.com/pages/about-us
5. Binotto C, Payne A (2017) The poetics of waste: contemporary fashion practice in the context of wastefulness. Fash Pract 9(1):5–29. https://doi.org/10.1080/17569370.2016.1226604
6. Black S (2013) Speed and distance/Ecology and waste. In: Black S (ed.) The sustainable fashion handbook. Thames & Hudson, pp 207−271
7. Brain E (2019) Footwear. Hypebeast. https://hypebeast.com/2019/8/converse-renew-denim-chuck-70-high-low-ox-beyond-retro-release-information
8. Bridgens B, Powell M, Farmer G, Walsh C, Reed E, Royapoor G, P., Hall, J., & Heidrick, O. (2018) Creative upcycling: Reconnecting people, materials and place through making. J Clean Prod 189:145–154. https://doi.org/10.1016/j.jclepro.2018.03.317
9. Brown S (2010) Eco fashion. Laurence King, London, UK
10. Burak C (2013)Craft and industry/Transparency and livelihood: Sustainability in the luxury industry. In: Black S (ed) The sustainable fashion handbook. Thames & Hudson, pp 188−189
11. Cassidy TD, Han SL (2013) Upcycling fashion for mass production. In: Gardetti MA, Torres AL (eds) Sustainability in fashion and textiles: Values, design, production and consumption. Greenleaf, pp 148−163
12. Chapman J (2005) Emotionally durable design: Objects, experiences, and empathy. Earthscan
13. Climate Emergency Declaration (n.d.) Call to declare a climate emergency. https://climateemergencydeclaration.org/
14. Clothes from Canada account for huge waste (2018) CBC: The National [Video]. https://www.cbc.ca/news/thenational/clothes-from-canada-account-for-huge-waste-cbc-marketplace-1.4494444
15. Copenhagen Fashion Summit (2018) Latest. https://copenhagenfashionsummit.com/latest/
16. Corbin JM, Strauss AL (1990) Basics of qualitative research: Grounded theory procedures and techniques. Sage
17. Creswell JW, Poth CN (2018) Qualitative inquiry & research design: choosing among five approaches, 4th edn. Sage

18. Crouch C, Pearce J (2012) Doing research in design. Bloomsbury Academic
19. Cuc S, Tripa S (2017) Can upcycling give Romanian's fashion industry an impulse? Ann Univer Oradea: Fascicle Textiles 18(1):187−192. https://tinyurl.com/ynempqhw
20. Cuc S, Tripa S (2018) Redesign and upcycling—A solution for the competitiveness of small and medium-sized enterprises in the clothing industry. Industria Textila 69(1):31−36. https://doi.org/10.35530/IT.069.01.1417
21. Dissanayake G, Sinha P (2015) An examination of the product development process for fashion remanufacturing. Resour Conserv Recycl 104:94–102. https://doi.org/10.1016/j.resconrec.2015.09.008
22. Ellen MacArthur Foundation (2017) A new textiles economy: Redesigning fashion's future. https://tinyurl.com/y4s79ht3
23. Exceptional Canadians: Twin sisters tackle textile waste with children's clothing company Nudnik (2018) The Globe and Mail [Video]. https://www.theglobeandmail.com/life/adv/article-twin-sisters-tackle-textile-waste-with-childrens-clothing-company/
24. Farrant L, Olsen SI, Wangel A (2010) Environmental benefits from reusing clothes. Int J Life Cycle Assessment 15(7):726–736. https://doi.org/10.1007/s11367-010-0197-y
25. Fletcher K (2014) Sustainable fashion and textiles: Design journeys, 2nd edn. Routledge
26. Global Fashion Agenda, Boston Consulting Group, & Sustainable Apparel Coalition (2019) Pulse of the fashion industry 2019 update. https://www.globalfashionagenda.com/pulse-2019-update/#
27. Han S, Tyler D, Apeagyei P (2015) Upcycling as a design strategy for product lifetime optimization and societal change [Paper presentation]. PLATE conference, Nottingham, England. https://tinyurl.com/m8rhugj3
28. Han SLC, Chan PYL, Venkatraman P, Apeagyei P, Cassidy T, Tyler DJ (2017) Standard versus upcycled fashion design and production. Fashion Practice 9(1):69−94. https://doi.org/10.1080/17569370.2016.1227146
29. Hsu C (2017) Selling products by selling brand purpose. J Brand Strategy 5(4):373–394
30. Janigo KA, Wu J (2015) Collaborative redesign of used clothes as a sustainable fashion solution and potential business opportunity. Fash Pract 7(1):75–98. https://doi.org/10.2752/175693815X14182200335736
31. Janigo KA, Wu J, DeLong M (2017) Redesigning fashion: an analysis and categorization of women's clothing upcycling behavior. Fash Pract 9(2):254–279. https://doi.org/10.1080/17569370.2017.1314114
32. Keith S, Silies M (2015) New life luxury: Upcycled Scottish heritage textiles. Int J Retail Distribut Manag 43(10/11):1051–1064. https://doi.org/10.1108/IJRDM-07-2014-0095
33. KinsuAtelier (2019) Ready-to-wear and ready-to-make brand. Etsy. https://www.etsy.com/ca/shop/KinsuAtelier?ref=seller-platform-mcnav
34. Kumar V (2013) 101 design methods: A structured approach for driving innovation in your organization. Wiley
35. Lunenburg FC, Irby BJ (2008) Writing a successful thesis or dissertation: Tips and strategies for students in the social and behavioral sciences. Corwin Press
36. Martin B, Hanington BM (2012) Universal methods of design: 100 ways to research complex problems, develop innovative ideas, and design effective solutions. Rockport
37. Maxwell JA (2013) Qualitative research design: An interactive approach, 3rd edn. Sage
38. Merriam SB (1998) Qualitative research and case study applications in education, Rev edn. Jossey-Bass
39. Minney S (2011) Naked fashion: The new sustainable fashion guide. New Internationalist
40. Moorhouse D, Moorhouse D (2017) Sustainable design: Circular economy in fashion and textiles. Des J 20:S1948–S1959. https://doi.org/10.1080/14606925.2017.1352713
41. Morgan LR, Birtwistle G (2009) An investigation of young fashion consumers' disposal habits. Int J Consum Stud 33(2):190–198. https://doi.org/10.1111/j.1470-6431.2009.00756.x
42. Muthu SS (2014) Assessing the environmental impact of textiles and the clothing supply chain. Woodhead.

43. Muthu, S. S. (2017). *Textiles and clothing sustainability: Recycled and upcycled textiles and fashion.* Springer.
44. Niinimaki K, Hassi L (2011) Emerging design strategies in sustainable production and consumption of textiles and clothing. J Clean Prod 19(16):1876–1883. https://doi.org/10.1016/j.jclepro.2011.04.020
45. Nudnik (2019) Clothing for their future. https://littlenudniks.com
46. O'Mahony M (2013) Key directions for textiles and sustainability in the coming decade. In: Black S (ed) The sustainable fashion handbook. Thames & Hudson, p 307
47. Paras MK, Curteza A (2018) Revisiting upcycling phenomena: a concept in clothing industry. Res J Text Appar 22(1):46–58. https://doi.org/10.1108/RJTA-03-2017-0011
48. Parker L, Maher S (2013) Hidden people: Workers in the garment supply chain. In: Black S (ed) The sustainable fashion handbook. Thames & Hudson, p 307
49. Preloved (n.d.) Our story. https://getpreloved.com/pages/about-us
50. Richardson J (1996) Vertical integration and rapid response in fashion apparel. Organ Sci 7(4):400–412. https://doi.org/10.1287/orsc.7.4.400
51. Rissanen T, McQuillan H (2016) Zero waste fashion design. Fairchild Books
52. Royer A, Charbonneau L, Bonn F (1988) Urbanization and Landsat MSS albedo change in the Windsor-Québec corridor since 1972. Int J Remote Sens 9(3):555–566. https://doi.org/10.1080/01431168808954875
53. Santiago E (2019) Converse adds sustainable denim materials to their Chuck Taylor 1970. Sneaker News. https://sneakernews.com/2019/08/12/converse-chuck-70-renew-denim-release-date/
54. Streit C, Davies IA (2013) Sustainability isn't sexy: an exploratory study into luxury fashion. In: Gardetti MA, Torres AL (eds) Sustainability in fashion and textiles: Values, design, production and consumption. Greenleaf, pp 207−222
55. Tranberg Hansen K (2013) Secondhand clothing and Africa: Global fashion influences, local dress agency, and policy issues. In: Black S, de la Haye A, Entwistle J, Rocamora A, Hoot RA, Thomas H (eds) The handbook of fashion studies. Bloomsbury Academic, pp 408−428
56. Triarchy (2019) About. https://triarchy.com/pages/about
57. Ulasewicz C, Baugh G (2013) Creating new from that which is discarded. In: Gardetti MA,Torres AL (eds) Sustainability in fashion and textiles: Values, design, production and consumption. Greenleaf, pp 148−163
58. UN Alliance for Sustainable Development (n.d.) Sustainable development goals. https://sustainabledevelopment.un.org/?menu=1300
59. Vadicherla T, Saravanan D Muthu Ram M, Suganya K (2017) Fashion renovation via upcycling. In: Muthu S (ed) Textiles and clothing sustainability: Textile science and clothing technology. Springer, Berlin, pp 1−54. https://doi.org/10.1007/978-981-10-2146-6_1
60. Voris R (1999)Shibori surface design. School Arts 98(9):6. https://www.thefreelibrary.com/Shibori+Surface+Design-a054492393
61. Weber S, Lynes J, Young SB (2017) Fashion interest as a driver for consumer textile waste management: Reuse, recycle or disposal. Int J Consum Stud 41(2):207–215. https://doi.org/10.1111/ijcs.12328
62. Wilson M (2016) When creative consumers go green: Understanding consumer upcycling. J Product Brand Manag 25(4):394–399. https://doi.org/10.1108/JPBM-09-2015-0972
63. Yin RK, Campbell DT (2018) Case study research and applications: design and methods, 6th edn. Sage

Analysis of Pakistani Textile Industry: Recommendations Towards Circular and Sustainable Production

Shahbaz Abbas and Anthony Halog

Abstract Pakistani textile industry has a considerable presence in the global market considering its unique fabric quality, design and pervasive cotton cultivation. Pakistan has a momentous position in the spinning of yarn and cotton exports in Asia. One of the reasons behind this is Pakistan's substantial agricultural productivity in growing cotton annually over million hectares of land after wheat and rice. However, ecological degradation, climate changes and significant population expansion have dragged the textile industry towards production decline as a consequence of resources unavailability particularly the cotton, energy and water. Apparently, this crisis is resulting into economic fluctuations, industrial shut downs and unemployment that are ought to be addressed. In order to leverage this cascade, ensuring the availability of resources for smooth industrial operations is imperative which can be optimized by circular and sustainable production. Circular economy is radically transforming the conventional production systems into sustainable ways of production for the optimal yield. The nexus approach analysed in this study is developed on circular economy concept that delivers a dual relationship mechanism between the significant processes of industrial systems of the textile industry. In addition, utilizing resources efficiently, significance of textiles along with associated hazards and enabling the industrial waste as a resource in the industrial process has been analysed. Moreover, the rational policy implications, necessary actions and recommendations towards a circular and sustainable Pakistani textile industry have been discussed in this chapter.

Keywords Pakistani textile industry · Sustainable production and consumption · Circular economy · Resource-energy-water nexus · Resources sustainability · Textiles' waste management

---

S. Abbas (✉)
College of Engineering, Chung Yuan Christian University, Taoyuan, Taiwan

A. Halog
School of Earth and Environmental Sciences, The University of Queensland, Brisbane St Lucia, QLD, Australia
e-mail: a.halog@uq.edu.au

S. S. Muthu (ed.), *Circular Economy*, Environmental Footprints and Eco-design of Products and Processes, https://doi.org/10.1007/978-981-16-3698-1_3

1 Introduction

In order to optimize sustainable development, manufacturing industries have a substantial contribution. The growth and sustainability of manufacturing industries are dependent upon the sustainability of resources consumption [1]. Therefore, industrialists and sustainability experts are seeking sustainable and alternate ways of production processes in order to address sustainability challenges, ensuring sustainable practices in each industrial process from the consumption of resources to utilizing waste. Particularly the incorporation of sustainability practices towards energy and water systems along with environmental conservation for promising sustainable industries [2].

Billions of people worldwide have already a limited access to the essential natural resources such as energy, water and land, and this may get worse with the population growth [3]. Eventually, the rapid population growth will influence widespread urbanization, climate change issues and extensive industrialization which may further enhance the resources sustainability challenges. Nevertheless, these challenges will affect industries, economy, agriculture and people [4].

The textile industry has a prestigious reputation in the global manufacturing industry, and the trending variations by several fashion icons have intensified the significance of this industry [5]. It is one of the most dynamic and largest industries in the global manufacturing industries; however, the rapidly growing population and challenges from climate change may foster the industry sustainability concerns. The projected challenges of climate change have increased the threat of resource scarcity in the coming decades, which may take this industry further into unfavourable situations, leading to unemployment and economic loss [6]. However, the quantity of required resources such as cotton, energy and water are already inadequate for effective textile industrial operations, and as a result, the industry is increasingly failing to meet production targets.

The chapter progresses to Sect. 2 with the discussion on the importance of circular economy and sustainability practices for the global textile industry followed by an overview of the Pakistani textile industry in Sect. 3. Significance of nexus for the textile has been addressed in Sect. 4. Effective circular and sustainable practices for the Pakistani textile industry have been presented in Sect. 5. Section 6 focuses on how circularity in the textile industry can bring a sustainable change. Sections 7 and 8 are about the proposed actions that can be implemented in the Pakistani textile industry for the sustainable ways of fibre, yarn and fabric production. Actions to be taken in order to manage the textile waste have been mentioned in the Sect. 9 followed by conclusions derived by the study in the Sect. 10.

2 Textile Industry: Circular Economy and Sustainability Significance

The textile industry includes all activities that derive or derive value from the design, manufacture, distribution, retail sale and consumption of textile products (or the provision of services provided by textile products), along with the supply of raw materials, including all activities related to textiles after the end of valuable service life. Apparently, the textile industry caters to all product's life stages such as raw material supply and after-use disposal, combined with the investments and regulations for the value creation and business models. Energy and raw materials are extensively required during all the intermediate processes including transportation and all other product manufacturing stages along with certain carbon emissions in the external environment.

During textile production, the industry starts its operations depending on the type of textile factory. This can either be the sourcing of natural agricultural materials and their subsequent processing to extract the fibre (e.g. cotton) or crude oil extraction, along with the development of chemicals from which synthetic fibres are made (e.g. polyester). As the textiles are frequently blends of natural and synthetic fibres, its production involves both natural materials and chemical processing. During the whole textile production, traditionally there is a linear representation of all the manufacturing processes starting from the production of raw material towards the end product along with the significant recycling, repairing and reuse of the potential materials.

For a circular economy, there is a transformation from a linear take–make–dispose process into a circular loop industrial system, in which the economic system has been sustained by utilizing all the materials as a resource and waste wherever possible in order to create value. In this sense, a circular industry makes a more appropriate representation. Nonetheless, a linear representation of the industry is more representative of the status quo and is convenient for indicating where stakeholders and impacts are located along the industry.

The industry has been influenced by all the stakeholders' activities including the people and other non-manufacturing activities that have been involved in the product manufacturing process. Therefore, the textile industry has not been limited only to the physical systems such as factories and agriculture farms, it also emphasizes on the industrial business model developed for the design and promotional techniques based on the consumers' choices and trends. During the textile production and consumption, there is a significant contribution of designing, advertising, retailing and marketing strategies as the non-manufacturing activities in the contemporary textile products.

2.1 *Measuring Circular Economy: Life Cycle Assessment Approach*

Findings on the socioeconomic and ecological effects of the textile sector have tended to focus on clothing and apparel. Box 1 provides the details of the methodology and data underpinning the study. The social risk results are taken from a social hotspots study, of which an overview is given in Box 4. These quantitative results are supplemented by the wider literature, especially for those environmental and social impacts identified as limitations in the LCA studies. The hotspots identified are therefore applicable to apparel, although the environmental profiles of household textiles (e.g. towels, linen, etc.) produced in a similar industry to apparel (i.e. spinning, knitting or weaving and textile production) are expected to be similar. However, the environmental profiles of industrial and technical textiles are potentially very different from that of global apparel presented in this section. Nonetheless, given that apparel and household textiles together account for 80% of global textile production, and that this high share is not expected to change [7], the insights presented here into the social and environmental impacts of textiles can be taken fairly as indicative of the textiles sector as a whole.

Box 1: Methodological overview of LCA studies informing the hotspots analysis

A 2018 research study conducted by FICCI for UNEP, Assessing Trade Barriers Mapping the Textile Industry, Opportunities and main hotspots identification globally and provides the quantitative basis for the hotspot analysis. Environmental and social life cycle assessment studies were carried out in this study, with the environmental LCA building on the Quantis 2018 study titled, “Measure Fashion: Global Environmental Impact of the Footwear and Apparel Industry”.

LCA is a life cycle assessment technique which evaluates a products’ or services’ environmental performance during its life cycle. The extraction of resources, and releases to air, water and soil are quantified at each life cycle stage, and the potential contribution of these extractions and releases to predetermined environmental impact categories is then assessed. LCA is therefore a good tool to provide a quantitative basis for a hotspot analysis.

The results of the FICCI and Quantis studies are based on the World Apparel Life Cycle Database (WALDB), with 2016 as the baseline year. The study considers multiple fibre materials, with the results reflecting the global apparel fibre mix in 2016. Data on global fibre production is taken from The Fiber Year 2017. No distinction is made between conventional fibre materials and more sustainable fibres, with all fibres assumed to be produced conventionally. The issue of micro plastics falls outside the scope of the studies.

In addition to the freshwater withdrawals covered by the Quantis study, the FICCI study assesses water impacts using the AWARE method. The AWARE scarcity of water footprint indicates the potential ways of depriving to water usage and its consumption to the ecosystem. The life cycle system diagram for the Quantis and FICCI global apparel LCAs. The following are considered in each of the stages:

Fibre Production

It involves the processes of extracting of fibres. Raw material transportation between the extracting and other processes, and includes the transportation between the yarn preparation and fibre production.

Production of Yarn

It covers the filament, the staple fibres and the yarn spinning. Various methods of spinning are taken into consideration (wet spinning and cotton spinning). Yarn to fabric production also require transportation.

Production of Fabric

It covers the process of yarn conversion into fabric by weaving and knitting. Knitting methods that are taken into consideration (circular and flat). Fabric production to the wet-processing (bleach, dye and finish) also requires transportation.

Dye and Finish Process

It covers the process to bleach and dye the fabric as well as fabric to finish the product. After this process, the dyed and finished fabric is required to be transported for assembling.

Assembling Process

It covers the process to cut and sew the finished fabric into garments and apparels.

Process of Distribution

It consists of apparel transportation to the market. The FICCI study includes the selling of garments to the end-users (retail).

Use

The use stage is not considered in the Quantis Measuring Fashion study, but is included in the FICCI study. Washing (at an assumed temperature of 30°C), ironing and drying of apparel products is included (with 50% of apparel products assumed to be ironed and dried electrically).

Disposal

It covers the process of collecting the end-of-life product for incinerating and landfill. It also requires transportation.

Consistent with the methodology of LCA, not only are the above processes making up the main life cycle stages taken into account, but all identifiable upstream inputs into them are included as well. For example, cotton farming includes the production of fertilizers.

A full description of data sources, assumptions, limitations and key uncertainties in the study can be found in the full Measuring Fashion Methodological Report [8]. Significant sources of uncertainty are due to data gaps in, amongst others, the proportion of fibres used in apparel, local water impacts of dyeing processes, and the geographical breakdown of manufacturing locations.

2.2 *Circular Economy Assessment and Nexus*

Over the time and space, the synergies and the trade-offs have been identified by the nexus concept [9–12]. The nexus concept has been extensively considered by enormous policy studies including circular and bio-based economy. There would be misleading modelling outcomes if the occurring trade-offs and synergies between natural resources including the energy and water would be ignored [13].

Following the reputed energy–water nexus concept, several other nexus has been developed by researchers in recent years in terms of managing sustainable development. Water–energy–food nexus along with their enhanced modelling tools can maximize the possible synergies and trade-offs synergies among the components of nexus in order to optimize resources sustainability [10]. In addition, understanding the interrelationship between energy, water consumption and land for crop production can be supported by the development of land–energy–water nexus involving appropriate agricultural policy measures [14]. In order to manage urban carbon emissions, energy and water, another significant energy–water–carbon nexus has been introduced [15]. On the other hand, resource–energy–environment nexus has been developed to assess the individual characteristics and interactions between the resources, energy and environment for China's steel and iron industry [12].

The above-mentioned nexuses have been modelled to address the reduction in carbon emissions and managing natural resources' sustainability challenges. Nevertheless, the sustainability of all manufacturing industries particularly the availability of energy, water and other raw materials is challenging because of substantial population growth, ecological degradation and industrial development [14]. In contrast, the optimization of resources, energy and water emphasizing the textile industry has not been significantly analysed.

3 Pakistani Textile Industry: A Case Study

3.1 Significance

Pakistani textile industry has a momentous position in the global textile industry. After China and India, Pakistan has the highest yarn spinning capacity, and it has been ranked at number eight in Asia as the largest cotton exporter [16]. In the Pakistani manufacturing industry, the maximum contribution to the economic development is dependent upon the textile industry. Based on its dependency on employing 40% of overall industrial workforce and being a labour-intensive industry, it is contributing more than a quarter of total industrial added value. Pakistani textile industry accounts for 50–60% of the national exports along with generating maximum foreign exchange earnings and its export orientation over other manufacturing industries. It has also competitive recognition in the international market based on the quality of yarn in weaving process, along with a significant domestic market [17]. Unfortunately, during the extreme weather conditions, the industry is suffering from on and off power outages leading to production decline along with the sustainability of raw material, energy and water, eventually affecting the sustainability of the textile industry [18].

The targeted factories to be analysed have been selected from one of the major industrial cities of Pakistan that is the Faisalabad city other than Multan, Lahore and Karachi. Certain factories in Faisalabad have been visited in order to obtain the data on energy, water and raw material consumption. Keeping in view the challenges of power outages and having the maximum number of wet-processing factories, this city has been selected over other major industrial cities of Pakistan. As a result of resources unavailability issues, particularly the energy and water, the stakeholders are transferring their investments to other non-cotton producing countries such as Vietnam and Bangladesh, keeping in view the suspended orders and more than fifty thousand units of energy deficit on average per day. The situation in meeting national textile production targets is getting worse over time with persistent power outages, as Faisalabad is an industrial hub for textile production. In addition to energy challenges, the industry is utilizing underground water for textile production consequently leading to water sustainability for the industry as well as for the citizens. Though, it is convenient to manage the raw material supply, particularly cotton in this city, as cotton has been significantly cultivated and harvested in the city premises and surrounding villages. Another challenge for this city is the environmental sustainability as a result of surface water contamination with the toxic textile wastewater containing chemicals and exposed carbon emissions in the air by burning the fossil fuels during power outages. Hence, environmental sustainability is also a major concern for the city other than finding out the solution for resource, energy and water sustainability associated with the textile industry.

3.2 Sustainability

The Pakistani government is doing tremendous efforts towards ecological and resources sustainability likewise other developing countries. Natural resources including natural gas, water, land and non-renewable energy resources such as fossil fuels have been depleting due to their excessive consumption, and it may get worse as a consequence of climate changes. Undoubtedly, economy, ecosystem, agriculture and industries have been affected by the climate changes.

Wet processing, weaving and spinning are the three different types of textile factories in Pakistan. Production of yarn from cotton has been performed by the spinning factories and transformation of yarn into grey fabric has been performed by the weaving factories. The conversion of grey fabric into printed, colourful and vibrant fabric has been performed by the wet-processing factories. Spinning and weaving factories are mostly dependent upon cotton as raw material and energy for yarn production; however, in addition to cotton and energy, wet processing factories consume water resources as well, leading towards having the longest value chain over spinning and weaving factories. Figure 1 illustrates the sequential flow of the main processes involved in textile manufacturing.

Pakistan is facing an imbalance for a couple of decades in the electricity supply and demand. There is an energy deficit of 6 GW as the demand for electricity is 23GW and the supply is almost 17GW [19]. This crisis may further accelerate in 2030 which may go up to 25GW as a result of a substantial spike in population

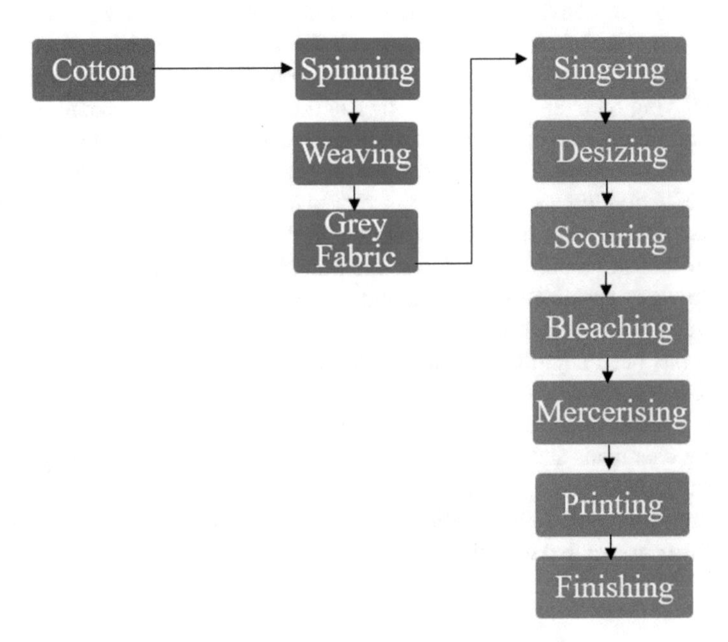

Fig. 1 Flow of textile production processes. *Source* Abbas et al. (2020)

statistics. In addition, the water resource depletion challenge in Pakistan is a serious concern as reported by the World Bank. On the other hand, the resources crisis is not only affecting the economy but the other significant sectors which are extensively dependent on water resources such as agriculture, production of hydroelectricity, domestic and industrial water consumption. Likewise, in order to achieve textile production targets and for the execution of smooth daily operations, an adequate supply of water and energy is needed.

One of the significant cash crops in Pakistan is the cotton crop and the textile industry is strongly relying on cotton as the raw material in order to achieve expansion and growth in the industry [20]. Land area covered for the cultivation of the cotton crop is around 2372 million hectares, and after wheat, it is the 2nd largest crop cultivated in Pakistan. Figure 2 represents the cotton production statistics by the economic survey of Pakistan 2018–2019 over a period of seven years, and it has been analysed that cotton production faced a decline from 11.946 million bales to 9.861 million bales resulting in a production downfall of 17.5% within two years [21]. The significant reason for this decline is that the areas of cultivation have been affected by the prolonged extreme weather conditions (colder and hotter weather with lesser rain) particularly because of extreme weather conditions. Consequently, sustainability of the textile industry will be affected aggressively due to the decline in cotton production.

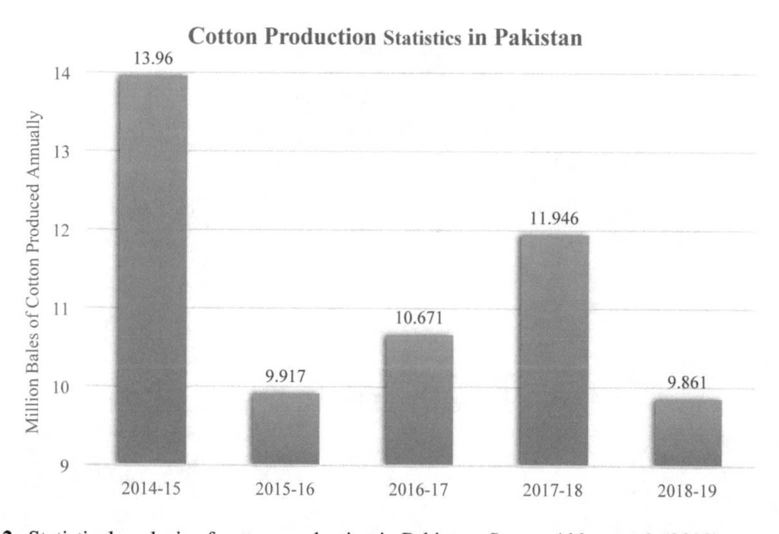

Fig. 2 Statistical analysis of cotton production in Pakistan. *Source* Abbas et al. (2020)

4 Resource–Energy–Water (REW) Nexus

In order to analyse the association between energy and water for the Pakistani textile industry along with the optimization of resource sustainability, system dynamics approach have been adopted for the textile factories. Cotton, energy and water are the resources that have been extensively consumed in the wet-processing textile factories, and the interactions between cotton, energy and water with a two-way relationship in a simplified form of nexus have been illustrated in Fig. 3.

Ecosystem represents the association between agricultural products and water relationship as resource–water interaction, and if resource waste such as cotton has been consumed as biomass energy, then it represents the interaction between resource–energy relationship [14]. During the production of yarn/fabric from cotton in the spinning and weaving processes, the energy–resource relationship shows the significant amount of energy consumed in these processes. During textile production, the consumption of energy for heating water by the boilers during the wet-processing is represented by the energy–water relationship [22, 23]. Cotton is an agricultural commodity, and the dependency of cotton crop for cotton production determines the relationship between water–resource. Heat recovery process from the textile wastewater can be adopted as a process of generating energy from the wastewater that clearly represents the relationship between water and energy [24, 25].

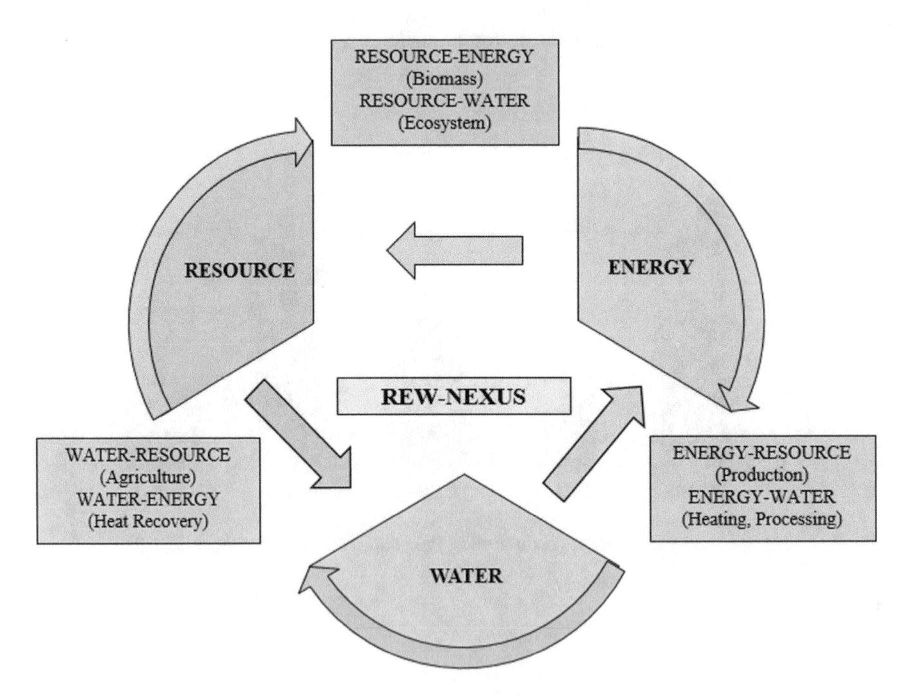

Fig. 3 Proposed REW nexus. *Source* Abbas et al. (2020)

Water, energy and raw material (resource) consumption during various textile processes can be calculated based on the nexus phenomenon in order to ascertain the parameters for the textile industry's sustainability. Nexus interactions also support to calculate and validate resources consumption during peak and as well as off-seasons of production. The patterns and trends of raw material, energy and water consumption can be estimated by different parameters and variables depending on the type and production of each textile factory.

Systems dynamics technique effectively supports to produce the interactions between nexus components such as resources, energy and water for the textile industry. Different complex environmental, energy and water systems have been analysed using system dynamics approach [4, 26–31]. In addition, nexus studies have adopted system dynamics approach for the integrated modelling [32]. For the case study in this chapter, complex and dynamics simulations have been developed using an industrial-based software Vensim in order to apply system dynamic technique in this study. From Fig. 4, the causal loops between different components of the proposed nexus and textile processes interactions have been illustrated. Textile industrial processes connecting all the resources, energy and water connections have been indicated in Fig. 4. Therefore, along with seasonal variations and temperature measurements, resources, energy and water parameters have been focused in this nexus analysis to make estimations for the textile industrial sustainability.

Energy consumed processes and other related processes have been represented in yellow boxes in Fig. 4. Higher production leads to higher energy consumption; therefore, the demand of daily textile production can determine the daily energy consumption. During the manufacturing process, wattage of the systems can estimate

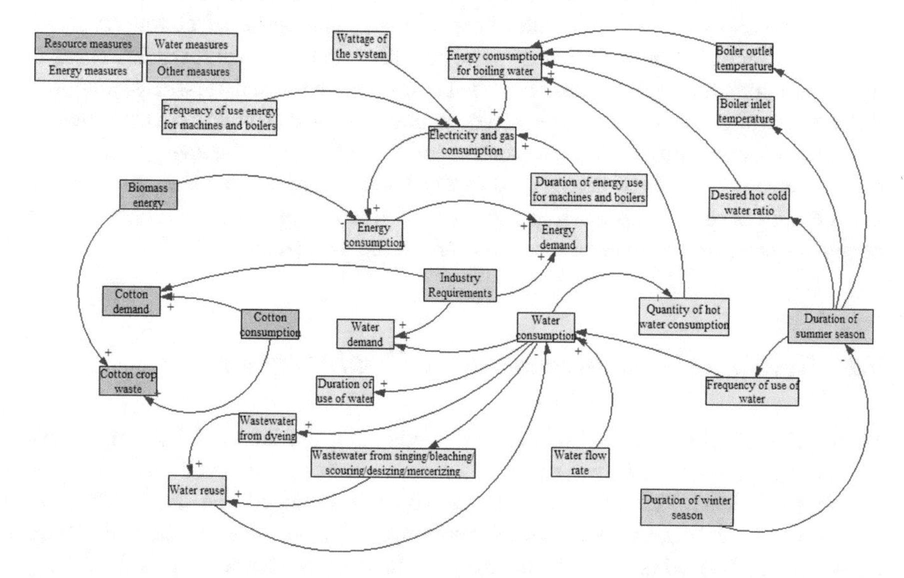

Fig. 4 Causal loops simulation based on system dynamics approach. *Source* Abbas et al. (2020)

the energy consumption. Furthermore, grid-supplied electricity and natural gas are the major energy sources of energy supply to the textile factory. On the other hand, coal as another form of energy has been consumed by the textile factories in order to manage the textile production during long hours of energy outage. Boilers are the systems that consume maximum amount of energy supply. Overall, during textile production, energy usage frequency, temperatures of the boilers, energy consumption hours per day and the induction motors for the mechanical transformation of cloth in the factory are the processes of major energy consumption. Although, few energy systems consume energy, however, they have not been significantly consuming the energy during the manufacturing process such as the air conditions, fans and lights. Some wet-processing textile factories in Faisalabad, spin and weave the cloth before wet-processing which can eventually influence the more energy requirements and consumption for a single factory but most of the factories involved in this study just perform wet-processing activities.

Consumption of water-based processes and other related parameters have been displayed by the blue boxes in Fig. 4. Daily production demand in the textile factory also determines the estimated water consumption. All the wet-processing involved in the systems during the singing, bleaching mercerizing, desizing, scouring and the printing of fabric significantly consume water resources for the boilers. Afterwards, this water has been trained to the municipal sewerage after the completion of all textile processes in the form of wastewater. The bleaching chemicals, dyes and other chemicals extensively used in the industrial production can be removed from the wastewater and reutilized in the system as a resource. All of these steps have been illustrated in the interaction between the components of water consumption in Fig. 4.

Figure 4 also displays a representation of seasonal variations and resource parameters, respectively, in the green and brown boxes. The waste of cotton crop is a significant source of biomass energy, and its utilization in the industrial process is dependent upon how much cotton has been consumed in the daily textile production. Industrial production targets have also been affected by seasonal variations such as the extreme summer and winter seasons as they directly affect the energy, water and cotton consumption. Overall, Fig. 4 expresses that there are rational interconnectivity between the nexus components, as represented in the form of causal loops, and the nexus has been further using the calculations in the next steps.

4.1 Resources' Consumption in the Textile Factory

During the peak seasons and on-demand production, production targets may vary depending upon the resource, energy and water consumption in each textile factory. In Table 1, at least 20,000 lb/d on average has been produced in a single factory in the city of Faisalabad, and the associated cotton, energy and water consumption have been estimated based on the systems dynamic model has been represented.

From the estimates of Table 1, it has been analysed that during winters, the consumption of all resources is higher as compared to the consumption in the

Table 1 Estimated calculations on cotton, energy and water consumption. *Source* Abbas et al. (2020)

Description	Unit	Estimate
Cotton consumption in water	Ton/h/year	1400
Cotton consumption in summer	Ton/h/year	1200
Energy consumption in winter	GWh/h/year	48
Energy consumption in summer	GWh/h/year	36
Water consumption in winter	M^3/h/year	8566
Water consumption in summer	M^3/h/year	4394

summers. There is a significant textile production in the winter season keeping in view the higher demand for women summer collection during the summer season, and the production has to be performed during the winter season to deliver the orders in the summers.

Farmers cultivate and harvest the cotton crop once a year, and they can only sell the cotton bales once a year to the cotton stockiest or directly to the factory managers. This once-a-year available raw material for the textile production along with fluctuations in the cotton crop production is a risk towards sustainable textile production. Seasons also affect the textile production when energy consumption is to be analysed. Natural gas and grid electricity are the only two energy resources available for the Pakistani textile factories as no sustainable forms of energy such as renewable energy-based systems have been installed in these factories. However, coal as an alternate form of fuel has been consumed which is a non-renewable and emission-based source of energy for the purpose of energy management during the power outages as mentioned previously in the model results.

On the other hand, as wet-processing factories are significantly dependent upon water consumption, therefore, water consumption has been estimated higher as compared to the energy and cotton consumption. Sources of water supply are the dams and groundwater; however, in order to facilitate the industrial, agricultural and domestic water consumption, the city may face unfavourable water availability challenges in the next years. Therefore, the industry is required to look for alternate and sustainable sources for water consumption.

5 Sustainability and Circular Economy Practices

Recognizing what is required to be done to achieve circularity and sustainability goals, first define what a circular and sustainable textile supply chain would look like. A sustainable textile industry is one that is resource-efficient and renewable resources-based, producing non-toxic, high-quality and affordable clothing services and products, while providing safe and secure livelihoods. A vision towards a sustainable textile industry outlined by UNEP has been summarized in Fig. 5. To achieve such an industry will require a shift in business model towards more circularity,

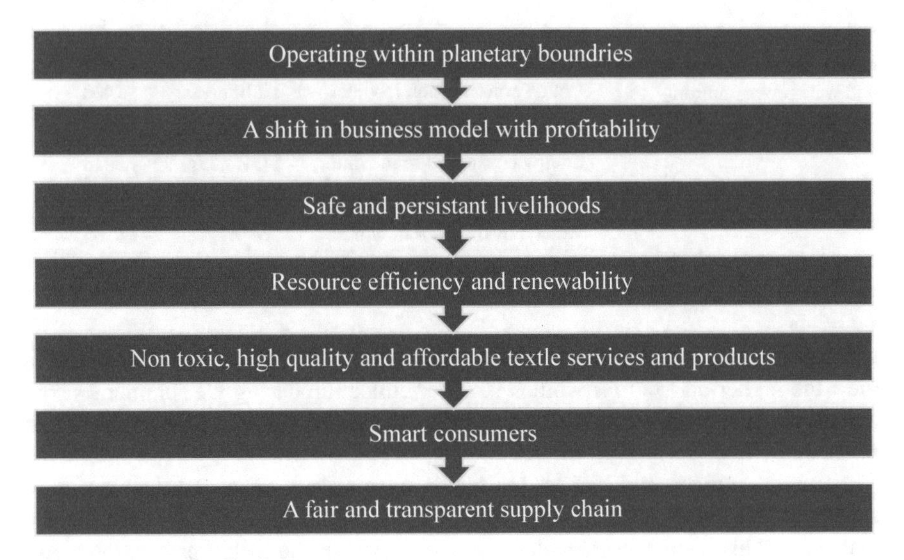

Fig. 5 Practices for a sustainable and circular textile industry proposed by UNEP

informed consumers and fair, transparent and traceable industry. Implicit in the definition of a sustainable textile industry is that it must operate within planetary boundaries, and that consumption cannot go unchecked, regardless of how efficient and circular the system is able to become.

Consumers (private consumption as well as public demand) have strong leverage in contributing to the circularity of textile products possibly by using their clothes for a longer time and not buying superfluous and unsustainable products. Motivating these more responsible consumer choices by setting new trends in the civil society associated with the textile industry is essential to accompany a sustainable change in practice. "User-to-user" value retention loops translate into three distinct circular processes:

Refuse: Saying no to purchase new clothes often. Another way is to discourage ecologically degrading complementary and unnecessary textile services or products particularly the unsustainable shopping bags. In addition, the design of the clothes, if found by the customers as environmentally unfavourable, should be rejected. Eventually, this behaviour can change the textile market trends towards sustainable and circular textiles.

Reduce: Saving the people and nature, consumers should be encouraged to modify their buying behaviour to minimize the waste impact and adapt their lifestyle towards buying less. If the consumers start prioritizing their purchasing decisions in buying lesser clothes, it can create a great impact. By doing this, consumers can not only save their money but they are also owing the environmental cost of textiles and adding value to the textile industry.

Re-use: Incorporating the old but wearable clothes can be reused by the consumers themselves, or they can sell their quality clothes on sale as second-hand textiles. In this way, consumers can perform their social responsibilities without any modification and without the involvement of intermediaries. In order to retain the product value, it is one of the alternate ways with a little investment. This step leaves an impression on the textile industry to produce more reliable and long-lasting products, when people will start prioritizing to buy such products which they can resell after a certain period of use, consequently, transferring the textile industry towards sustainable production and consumption.

Repair: This refers to the fixing of a specified fault in a product which would otherwise be considered as waste, in order to make the textile product fully functional for use of its originally intended purpose—thus extending its product lifetime. A user sends the product for repair, to a business intermediary, through the retailer or directly to repair shops. The textile product comes back to its original user or to a new one. Repair can also be considered as a service to users. Producers, in cooperation with other industry stakeholders (designers, producers, retailers, waste handlers, recyclers, raw material producers, etc.) need to work together to ensure discarded textile goods and components are not lost to disposal processes but are instead used as materials in other product systems. It translates into the following two circular processes:

Repurpose: A new distinct life cycle of a textile can be made possible by transforming the damaged or unwearable textile into a useful product with certain modifications. Doing this sustains the product's value along with economic benefits to some extent. Repurposing brings design singularity with a different function, along with cost reductions and textile waste.

Recycle: Bringing back the textile into the economic system along with waste minimization is recycling. In this processing, the product is converted into some new useful by completely reprocessing of material including new design, new texture and functionality with little to major changes. In order, to implement this recycling of textiles, an intensive infrastructure is required which demand investment and capital, however, reducing the waste.

6 Why Circularity Is Important? Environmental Impacts of Traditional Textile Production

6.1 Impact on Water Resources

The Pakistani textile industry consumes billions of litres of water per year [8].

Textile production stages that are significant consumers of water are raw material production, bleaching, dyeing and finishing in textile production, and use (laundering).

High water use in fibre production is due to the high levels of water required in growing cotton. Other natural fibres that do not require irrigation make a much lower contribution to industrial water use, while synthetic fibres require relatively little water in their manufacture.

The impact that water use has on water availability for human and industrial purposes and for ecosystem services varies from country to country, as each geographical region experiences different degrees of water scarcity; however, it depends on the number of competing users and the availability of freshwater. When weighted for country-level water scarcity, raw material production (cotton growing) makes the highest involvement in the apparel industry in terms of water scarcity footprint, and then the yarn production. The distribution of water impact across the industry is strongly influenced by the proportion of cotton in the apparel fibre mix. The **FICCI** hotspots study (see Box 1) also looks at water impact in terms of the countries most affected. The water scarcity footprint accounts for China is 34% share of the global textile industry. This high percentage is because China both grows cotton and has a significant contribution in apparel production. On the other hand, the countries with the next highest shares of the water scarcity footprint of global apparel are the USA (5%) and India (12%). These three countries have more land, resources and economy as compared to Pakistan particularly China and the USA.

The manufacture and use of textile products are not only associated with consumption of large volumes of water but the chemicals and detergents used in manufacturing processes and in washing textiles pollute natural waterways when effluents are released without sufficient treatment. Thus, the textile sector has a significantly larger impact on water scarcity than direct water use alone, by polluting water and rendering it unfit for other uses. A further impact that the textile sector has on water quality is the release of microfibres. Microfibres are the tiny strands of staple fibres or filaments that have been found just about everywhere that studies have tested for them, from bottled drinking water to Arctic ice. The prevalence of microfibres in wastewater and wastewater treatment sludge, together with the relationship observed between the abundance of microfibres in shoreline sediments and human population density, has led to laundering of textiles being identified as a major source of microfibers [33]. The emerging issue of microplastics8 arising from the textile sector is explored further in Box 2.

Box 2: Microfibers in textile production

During the textile production to the end-of-life management, textiles contribute to environmental degradation also by expelling microfibres in the air, though only limited evidence is there to find out the potential hotspots of microfibre release. Research on microfibre release from textiles has tended to focus on the parameters affecting release, such as washing machine type, wash duration, wash temperature and detergent use, as well as on the potential for different kinds of fabrics to shed [34–36].

Along with the use phase, textile processing is likely to be a significant source of microfibres. Roos et al. [37] identified production practices that reduce shedding in garment production, and found no evidence to suggest as compared to vigin fibres, the fabrics produced by the recycled fibres expel more microfiber. High quantities of microfibres were found to be released from a textile production wastewater treatment plant even after 95% of microfibres had been removed, with the high volumes of effluent released from textile processing translating into significant quantities of microfibres released even when their concentration in the effluent was low [38]. Little evidence of the potential for textiles to release microfibres at their end-of-life is available in the literature, although landfills have been identified as a potential source of airborne microfibres [39].

Recent research indicates a higher presence of microfibres of natural and semi-synthetic origin than previously appreciated [40, 41]. Cellulose-derived microfibres have been found in high concentrations in a number of different environments [33]. However, while natural fibres are biodegradable, which potentially lessens the ecological degradation, however, the dangers associated have not been properly analyzed so far [41], for example, in terms of the time taken to biodegrade in the marine environment and the release of chemicals contained in the fibre. Despite how ubiquitous microplastics are in the environment, the mechanisms causing their ecological impacts are poorly understood. This is in part due to the multifaceted nature of the potential impacts, with evidence of individual or integrated chemical, biological and physical, chemical processes [33].

During ingestion, there is an occurrence of major physical impact, the effects that have been relatively well documented for marine organisms, but less so for terrestrial organisms. Fire retardant or dyes are the leached toxic chemicals that chemically impact the environment by microfibres [42], while biological & environmental impacts include the potential for microfibres to carry POPs and provide a habitat for pathogenic bacteria, thereby enabling the spread of such disease-causing bacteria to new locations and habitats [43].

6.2 Impact on Land Use

Land use is one of the main drivers of worldwide loss in biodiversity. It is responsible for 2/3 of the global terrestrial area having declined beyond a safe level in terms of biodiversity intactness within planetary boundaries [44]. Land use associated with global apparel is strongly weighted towards the fibre production stage. Synthetic fibres have only a small land footprint. The contribution to land use of the other industry stages is indirect, in that it relates to the land associated with producing the energy used in manufacturing and laundering textiles. Whether individual countries have land-use profiles similar to the global apparel profile will therefore depend on the particular energy mix of the country, and especially the degree of biomass in the energy mix.

The dominance of fibre production in the industry land footprint is even more remarkable considering that in 2016 (the baseline year of the analysis shown in Fig. 12) natural fibres made up about one-third of global fibre production. While 2.5% of the world's arable land has been reserved for cotton cultivation. Other natural fibres also have high land footprints, with wool at the top end of the scale, requiring only 1 hectare per ton for cotton and 278 hectares per ton of rest of fibres (although wool is in many cases a by-product of meat production, with grazing often taking place on land that is not suitable for growing crops, complicates the direct attribution of land use to wool) [45]. Regenerated or cellulosic fibres, such as viscose, modal and lyocell, have smaller land footprints than other fibres produced from agricultural sources. However, given the steady increase in demand for these fabrics and the fact that over 140 million trees were used for making viscose in 2018, it is of paramount importance to ensure that wood is not sourced from ancient and endangered forests or other controversial sources [46].

6.3 Impact on Ecosystem

The high impact of cotton cultivation on ecosystem quality is due to land use (habitat loss), water use, soil degradation and high use of agricultural chemicals. Global cotton cultivation is estimated to require 200 thousand tonnes of pesticides and 8 million tonnes of fertilizers per year [47]. Cotton is a water-intensive crop, grown predominantly in dry regions. Extensive and/or poor irrigation practices severely impact regional freshwater resources, potentially depleting surface or groundwater bodies, and affecting river catchments and wetlands downstream of water extractions. Furthermore, the agrichemicals used in growing cotton pollute freshwater ecosystems with excessive nutrients, salts and pesticides [

Textile production is a chemical-intensive sector, using and releasing hazardous chemicals with significant human health and environmental impacts (see Box 3). Toxic chemicals are used and released all along the supply chain from the production of the raw material to the finishing of the articles and in waste management (chemicals can leach out as textiles degrade in landfills, while incineration can lead to harmful emissions) [49]. Ecosystem impacts are generally underestimated in LCA studies on textiles due to the lack of availability of types of chemicals and their required measurements for analysis along with the analysis of environmental effects from these chemicals by using the models based on LCA [37]. However, when these are included, such as in [25], and when the focus is only on direct emissions (i.e. excluding indirect toxic emissions associated with energy production and fossil fuels) the bleaching/dyeing and finishing stage of textile production is a clear hotspot in terms of ecotoxicity impact. For the six garment types considered by [25], the high freshwater ecotoxicity impact of wet treatment was found to be caused mainly as a result of significant usage of toxic chemicals in the industry.

Energy-intensive stages in the textile industry are shown by LCA studies to be hotspots in terms of their impacts on ecosystem quality [25]. This is because mining and emissions associated with burning fossil fuels, particularly coal, have high ecosystem impacts. The high use of fossil energy in textile finishing and the electricity consumed in the use phase results in these industry stages being hotspots for energy-related impacts on ecosystem quality.

Box 3: Chemicals' use in wet processing of textiles

The textile industry is notorious for its impact on water systems. Despite this notoriety, surprisingly little data exists on the scale of water pollution from textile processing, and the often cited claim that treatment and dyeing of textiles is about 20% of industrial water pollution. Producing textiles requires a considerable array of toxins including biocides to reduce bacteria or mold growth and dirtand water repellents [49, 50].

On average, producing 1 kg of textiles requires 0.58 kg of various chemicals [47]. Kant [51] estimates that in excess of 8,000 chemicals are used in the various textile manufacturing processes. A study by the Swedish Chemicals Agency identified approximately 3,500 substances being used in textile production [52]. Of the 2,450 substances able to be analysed (the rest were not analysed due to confidentiality), 750 were found to be hazardous to human health, with 299 more riskier for the public, i.e. substances intentionally added and expected to remain in the finished articles at relatively high concentrations. 440 substances were found to be environmentally hazardous, with 135 of these functional substances of high potential risk to the environment [52].

China is the largest consumer of textile chemicals, accounting for 42% of global consumption. Of China's textile chemical consumption, 41% are surfactants (including dye additives, antistatic agents and softeners), 24% are sizing chemicals and 13% are lubricants [50]. Many of the chemicals used in textile production are known to have adverse health and environmental impacts. Hazardous chemicals found in effluents from textile processing facilities include some known to cause cancer and disrupt hormonal systems in humans and animals. Toxic chemicals, such as alkylphenols and per fluorinated compounds (PFCs) are particularly problematic as they cannot be removed by wastewater treatment plants. Flame retardants, including brominated and chlorinated organic compounds, are another particularly hazardous class of chemicals used in the production of some textiles. Many dyes contain heavy metals, such as lead, cadmium, mercury and chromium (VI), known to be highly toxic due to their irreversible bio accumulative effects, whilst azo dyes contain carcinogenic amines [53].

7 Proposed Actions for Sustainable Fibre Production in Pakistan

7.1 Natural Fibre

The production of natural fibres is a particular hotspot in terms of ecosystem quality and water scarcity impacts, especially cotton, with its high use of water, land and agrichemicals. Actions required, therefore, are to develop and roll out farming practices that reduce these environmental impacts (water, land and chemical use). Cotton cultivation is also associated with high social risks, including injuries and exposure to toxins and hazards, low wages, forced labour and child labour, gender inequality, corruption and fragility in the legal system. Actions required are government regulations against the use of toxic substances and harmful labour practices and better enforcement of legislation protecting workers' rights and the environment (where this already exists).

Many cotton initiatives address both the environmental and socio-economic impacts of cotton farming, although some have a particular focus, such as growing organic cotton, increasing water efficiency or fair trade such as better cotton initiative (BCI). BCI works with equivalent cotton standards in Australia, Brazil and multiple African countries. Alongside improving cotton production practices, one critical aspect of many of the cotton initiatives is to develop supply chains that connect sustainable cotton to brands, retailers and manufacturers, including connecting intermediary partners across the supply chain (e.g. traders and processors). Traceability is therefore a central component of many cotton initiatives.

Recognition of the large number of initiatives working towards more sustainable cotton led to the formation of the Cotton 2040 initiative, which brings sustainable cotton standards together, along with industry initiatives and leading brands and retailers. One major output of the Cotton 2040 initiative has been the Cotton Up guide, which provides a practical resource for brands and retailers wanting to source sustainable cotton.

Despite good progress in initiatives addressing the social risks and environmental impacts of cotton production, the reach of such initiatives needs to be extended; after a decade of operation, "better cotton" (as defined by the Better Cotton Initiative) accounts for 19% of global cotton, with a reach of two million farmers in 21 countries (Better Cotton Initiative, 2019). Growing this percentage and increasing the global coverage will require policy support from governments in cotton-growing countries, particularly in enforcing (or implementing) environmental protection laws and protecting workers' rights. Increasing the share of "better cotton" will also require the increased engagement of consumers and brands/retailers to create demand for sustainable cotton. Governments also have strong leverage in creating demand for sustainable cotton through implementing sustainable public procurement requirements for cotton. Furthermore, delivering on increased demand for sustainable cotton will require traceability in textile supply chains to move from being a niche "nice to have" to a mainstream requirement for textile products.

Other natural fibres, such as jute, coir, flax, sisal, hemp, ramie, kapok and kenaf, show potential as sustainable alternatives to cotton. However, a recent study found that none of these fibres has the technical feasibility to match the comfort and technical properties of cotton [37, 54]. Furthermore, there are insufficient studies to determine whether these alternatives are always preferable environmentally [25, 37]. The few life cycle assessments that have been conducted show a wide range of performance, largely due to methodological differences in the studies. Thus, further standardization and research are required to establish the potential of these fibres, and if they prove to be more sustainable, actions to grow their market share from their current low levels should be pursued.

7.2 Synthetic Fibre

Producing synthetic fibres from secondary materials has been a successful area of innovation, with the development of a number of fibres and fabrics produced from waste materials. However, this has largely been motivated by the plastic litter crisis, i.e. fibre produced from plastic bottles and ocean plastics, with the recycling of synthetic textiles at end-of-life still at very low levels. Systemic actions avoiding the consumption of non-renewable resources will ultimately be the most effective in addressing the impacts of synthetic textiles, such as innovative circular business models that extend product lifetimes and promote the reuse and repair/repurposing of textiles.

Synthetic fibres have been steadily increasing their share of global fibre production. A swing back to natural fibres will ameliorate the impacts of synthetic fibre production, along with the release of microfibres (microplastics) associated with the use of synthetic fabrics. However, the production of natural and regenerated fibres is also associated with social and ecological effects and any material swapping should be carefully assessed across the whole life cycle, including consideration of total volumes. A recent review of fibres finds that the best environmental outcomes are achieved when the functional properties of the fibre are considered along with an environmentally appropriate product life cycle (i.e. by taking into account the use phase and end-of-life management and not only the production of the fibre) [25, 37]. Furthermore, the review finds that there are no clear "winners" when it comes to sustainable fibres. Rather, the range of environmental performance within each fibre type (representing differences in manufacturing practices) is often larger than the differences between fibre types, thereby making it impossible to draw clear conclusions around their relative performance.

The development of new innovative fibres is needed, especially those that can be used for longer or reused and/or those that do not shed microplastics. However, life cycle assessment or impact studies are required to ensure there are no unintended consequences with new materials. While there are a number of new fibres reaching the market that claim to be more sustainable, there is often no data available to support such claims, and in general, no evident data has been found on fibres in this regard [25, 37]. Thus, alongside the development of new fibres, actions are required to increase the availability of life cycle data on the production of fibres, as well as on the production, use and end-of-life of textiles made from them.

7.3 Outcomes of Traditional Textile Production

The considerable under-utilization of clothing and the very low rates of repurposing and recycling of textiles after use represent a considerable loss of material value. Value loss occurs through textile products not being kept in service for as long as they could be, not being resold or repurposed when consumers discard them still in good condition—or not being sold in the first place, and being landfilled or incinerated rather than remanufactured or recycled when they reach material end-of-life. Globally, the cost of wearable textile thrown by the consumers is estimated at $460 billion17.

A study on textile fibres' material flows by the Ellen MacArthur Foundation found that just 13% of the fibre input for clothing is recycled. Less than 1% of this is closed-loop recycling, i.e. fibre recycled back into clothing, rather than into lower value uses, such as cleaning cloths and insulation. This is estimated to equate to an annual material value loss of more than $100 billion [47]. It is however worth noting that one conclusion of The Circular Fibre analysis is that data consolidations and improved reporting are required worldwide, given the lack of knowledge of what happens to textiles at end-of-life [47].

One of the reviews on textile recycling and reuse found that reuse is always more beneficial than recycling, and that, in general, as compared to landfilling and incineration, textile reuse and recycling reduce environmental impacts compared to incineration and landfilling [25]. Smaller recycling loops are more environmentally beneficial than larger loops. That is, recycling back to fabric has the potential to avoid both the production of raw materials and the subsequent fibre, yarn and fabric production processes, while recycling back to fibre only avoids the production of raw materials. For those impacts where textile production accounts for the majority of the impact, such as climate impact, recycling back to fibre can have relatively low mitigation potential (if any at all, if the recycling process itself has high-energy inputs). However, recycling cotton fabric back to fibre can potentially reduce the water footprint by 90%, since raw material production accounts for a significant majority of the water impact. Nonetheless, while fabric recycling can potentially mitigate more impacts than recycling back to fibre, fabric recycling may often be unfeasible due to the material being too worn out or the difficulty of finding a suitable end-use [37].

Furthermore, the type of fibre being recycled makes a difference. Recycling cotton has the potential to mitigate freshwater depletion and the use of pesticides and fertilizers (along with their impacts), while recycling polyester fibres has the potential to mitigate fossil resource depletion and climate impact [37]. Finally, it is important to note that closed-loop recycling (i.e. recycling textiles back into textiles) is not automatically "better" than open-loop recycling. There are cases when recycled textile materials hold a much higher economic value in another industry sector [37]. One example is using low-grade recycled textile fibres to reinforce thermoplastic biocomposites for the automotive industry.

Despite the clear environmental benefits of extending the life of clothing, the reuse of textiles can lead to both positive and negative socio-economic impacts. A growing movement to recycle and reuse textiles, particularly in the European market, has seen used clothing collected and exported overseas. The sorting and trading of used clothing create business opportunities and employment in both the exporting and importing countries while generating government revenue through tax and providing access to affordable clothing. The export of 12,000 tonnes of Nordic textiles to Africa is estimated to support more than 10,000 market sellers and their families [55]. However, there are also potentially negative effects, with the importation of used clothing putting local textile producers out of business and flooding landfill sites with waste textiles in countries that typically do not have the waste management facilities to deal with them [55–57]. However, studies are in development that look more deeply into the implications of the export of used textiles [55].

8 Actions in Sustainable Yarn and Fabric Production in Pakistan

The environmental impacts of yarn and fabric production stem primarily from the use of fossil-based electricity in their manufacturing processes (spinning and knitting or weaving). While they are not a hotspot in the industry, in that yarn and fabric production does not show the highest impacts of all the textile industry stages, their high energy consumption nonetheless warrants attention. Increasing energy efficiency in manufacturing and a shift to renewable energy are actions required to decrease the environmental impacts of yarn and fabric production. Yarn and fabric production also have high social risks, most notably poor working conditions, remuneration below the minimum wage, forced labour and poor health and safety standards. Requirements for greater transparency and traceability in textile manufacturing chains and enforcement (or implementation) of legislation protecting workers' rights are actions that are needed. Many of the platforms and programmes seeking to advance sustainability in the textile sector include both the socio-economic and environmental dimensions of sustainability, and/or are focused on addressing the sustainable development goals (SDGs). Many of the initiatives are multi-stakeholder with strong industry participation, particularly of large brands and retailers.

Initiatives seeking to improve transparency and traceability in textile supply chains are important enablers of sustainability initiatives in the textile sector (see Box 4). This is especially relevant for yarn and fabric production, since the higher up the industry, the more difficult traceability becomes. Reputational risks to brands and retailers, especially of human rights abuses in their supply chain, are a strong driver of improved labour practices, yet many brands are unable to trace their supply chains beyond assembly. This has seen the development of a number of sustainability standards with traceability and transparency as a core aspect. New technologies, such as blockchain, present opportunities for supply chain traceability, potentially able to provide consumers with garment-specific sustainability information.

Many initiatives have seen steady improvements being made in the textile industry, although it is recognized that the industry still has far to go [58]. Advances are primarily being made by large players and the premium segment, with small producers, especially in the entry-level price segment, making little headway. This is particularly concerning given the relatively large share of the global market made up of small producers and producers in the entry-level price segment. Actions are thus needed to ensure that improvements reach all players in the industry.

Actions relevant to driving sustainability changes in yarn and fabric manufacturing are relevant across all textile manufacturing stages and include disseminating knowledge about sustainable alternatives, cleaner production, resource efficiency and renewable energy, and building the skills and capacity needed to implement sustainable changes. Further actions include removing the entry barriers for smaller players especially through, among others, harmonizing guidelines and standards, devising incentives for companies to change to sustainable alternatives, and creating cooperation, funding and collaboration across the industry. There is a need to deepen and extend existing alliances for the implementation of sustainable practices, and for global coordination of initiatives and efforts. One of such initiatives targeting the need for coordination as it relates to actions to address climate change is by the fashion council of the UN. The Charter sets out the vision for the fashion and clothing industry of achieving net-zero emissions by 2050, with signatories indicating their commitment to support the goals of the Paris Agreement in limiting global temperature rise to well below two degrees Celsius above pre-industrial levels.

The Fashion Industry Charter for Climate Action does not constitute a new formal initiative or registered organization, but rather the work is carried out by the signatories with facilitation and coordination from UN Climate Change.

Change cannot be expected to come from within the industry alone, and governments and consumers have a critical role to play [59]. This includes regulators creating a legislative environment in which companies are accountable and driven to take action against poor labour and environmental practices. Governments play their significant role by promoting and implementing guidelines and policies for the businesses and enterprises set by OECD and, more broadly, promoting responsible business conduct. Responsible business conduct is highly relevant for policy makers wishing to attract quality investment while ensuring that business activity in their countries contributes to broader value creation and sustainable development.

Governments can promote and enable responsible business conduct through a number of actions: regulating (establishing and enforcing an adequate legal framework that protects the public interest and monitors business compliance); facilitating (clearly communicating expectations on what constitutes responsible business conduct, and providing guidance with respect to specific practices); working with stakeholders in the business community, worker organizations, civil society and the general public, and working across internal government structures as well as with other governments to create synergies and establish coherence with regard to responsible business conduct; demonstrating support for best practices and acting responsibly in the context of the government's role as an economic actor.

Capacity building within governments is required to allow better enforcement of regulations and ensure that, at a minimum, companies comply with national laws protecting workers' rights and the environment. Further, there is a need for a policy environment and infrastructure that make the transformation and implementation of relevant technologies possible. Finally, consumers need to be educated and provided with reliable information in order to be empowered to make ethical purchases.

Box 4: Advancing sustainability practices in the Pakistani textile industry

Transparency and traceability are critical enabling factors in practically all initiatives to improve the sustainability of textile products socially and environmentally. They form a key metric in textile product labels, standards, benchmarks, voluntary certifications, pledges and agreements. There are also a number of standards that are not specific to textiles, for example SA8000 (social accountability) and CDP reporting (carbon disclosure). The International Trade Centre (ITC) Standards Map24 provides an objective benchmark of different labels/schemes according to product/ service, producing country and market covered. For individual companies, traceability to key mid-stream suppliers, also called control points, who may have greater visibility and leverage over their own suppliers and business relationships further up the supply chain, can be an option. Despite being a requirement in standards, achieving transparency and traceability presents a considerable challenge in textile industry. The United Nations Economic Commission for Europe (UNECE) and the ITC with its Centre for Trade Facilitation and e-Business (UN/CEFACT) are conducting a project to address these challenges in the garment and footwear sector (from raw material production to retail).

The overall objective of the project is to strengthen sustainable consumption and production patterns through the development and implementation of an international Framework Initiative and a Transparency and Traceability Tool. A pilot project launched in January 2020 is to implement block chain technology for traceability and due diligence in the cotton industry from field to distribution. With industry partners in Egypt and Europe, the pilot will demonstrate end-to-end traceability of a product, and test cost-efficiency, scalability and transferability. Consumers can only take more sustainable decisions if they are provided with accurate and reliable information. UNEP and ITC's "Guidelines for providing product sustainability information" aim to help producers make reliable claims about their products' sustainability performance and thus enable informed consumer choices. They have been tested in various sectors, including textiles, and a number of tools and case studies are available to stakeholders wanting to improve the way they communicate textile sustainability.

Knowing the composition (fibre mix) and chemical content of material for recycling is critical. Thus,traceability is also very relevant for increasing material recovery after use. One initiative with potential for textiles is "product passports"—a set of information about the components and materials contained in a product25. Also working on this topic is the "Green Markets and Global Industry" work stream of UNEP's Environment and Trade Hub,which aims to enhance the design and uptake of sustainability standards and to facilitate market access for sustainably produced and certified products. As part of the Partnership for Action on Green Economy (PAGE), the Environment and Trade Hub has provided training in one of China's leading regions for textile production and export. The Hub also offers methodologies and resources on sustainability standards which have relevance to the textile sector, including a "Guide for the Assessment of the Costs and Benefits of Sustainability Certification", a handbook on "Trade and Green Economy", and an analysis on "Green Economy and Trade – Trends, Challenges and Opportunities".

9 Actions for Managing Textile Products Waste in Pakistan

It is important to distinguish between material end-of-life and product end-of-life, where the latter is perhaps better called "after-use" or "end-of-first use". This distinction is necessary to understand that reuse, repair/repurposing, recycling to fabric and recycling to fibre are all part of the solution of a sustainable and circular textile system. Prolonging the use of textiles is by far the most important action when it comes to reducing environmental impacts. However, ultimately the material will reach the end-of-life, preferably after a long use life and a large number of reuse and repair/repurposing cycles.

Material recycling then has an important role to play, although for recycling to be part of a sustainable textile system, the energy, water and chemicals used in the collection, sorting and recycling textiles must be less than that used to produce them, and supporting infrastructure must be in place [37]. Actions are thus required to further develop emerging recycling technologies, and to put in place sufficient policy and infrastructure support. Necessary steps for the textiles long term consumption and reprocessing from their very low levels are an important part of achieving a circular and sustainable textile system, but it should be recognized that, by themselves, these will never provide the solution. Furthermore, overarching actions, such as those to increase transparency (see Box 4) and for materials to be "toxic free", are prerequisites for increased material recovery at end-of-life. It is imperative to know the chemical content of recycled textiles as this determines the application in which the recycled material can be used. Despite legislation restricting hazardous chemicals in some countries, legislation and/or enforcement is still largely lacking in the countries dominating textile production.

Furthermore, certain chemicals are allowed because of the value and function they bring to the final product, or because the exposure levels are estimated to be low. High-value recycling opportunities thus remain unattainable because of the need for expensive technologies to remove toxins. To be viable, large-scale textile recycling facilities require consistent feedstock material, and this is currently hampered by a lack of traceability, insufficient or absent fibre labelling legislation and mixed materials (with many fabrics and yarns being blends of natural and synthetic fibres and the latter blends of different polymers) [37]. Textile recycling will thus remain at its current low levels unless actions are taken to ensure recyclers can be confident regarding the source, composition and chemical content of their feedstock.

While the focus tends to be on consumer actions, it is also important for companies to take additional actions to increase reuse and recovery within the textile manufacturing stages, for example, recovery of offcuts in garment assembly and rejects in textile production. These actions fall broadly under increasing resource efficiency in manufacturing. One regulatory approach to avoid waste altogether in the context of unsold goods was recently introduced in France. Significant regulations for the manufacturers' responsibilities, recycling and recycled content targets and taxes on landfill, coupled with innovation and consumer education, are needed to shift recycling from being a niche activity to a core component of brands' and retailers' business.

Furthermore, innovative recycling technologies can help to close material loops at textile end-of-life. France introduced EPR rules for textiles in 2008, and in Sweden, the government is working towards implementing EPR for textiles by 2025 [37]. However, none of the advanced recycling technologies has yet reached market maturity [60]. Thus, further financial and technical support for textile recycling technologies is needed. Overall, it is important to consider potential trade-offs regarding the environmental and socioeconomic impacts associated with increased textile reprocessing. For example, centralizing advanced textile recycling technologies might require the shipping of textiles (with the associated climate pact) and shifting textile manufacturing locations (causing socio-economic impacts). Furthermore, recycling processes require resources and energy and can themselves impact water quality. Life cycle-based studies should thus be undertaken of proposed actions and technologies to ensure these offer environmental and socioeconomic benefits to the textile system as a whole.

10 Possible Drivers of Circular Economy

The proliferation of circularity and sustainable production is dependent upon certain drivers and those major required priority driver areas of action are:

- Stronger governance and policies.
- Collaboration and financing.
- Change in consumption habits.

10.1 Stronger Governance and Policies

Governments are an essential driver for change, and gaps in policy and legislation are hampering the move to a more sustainable and circular textile industry. While actors in the textile industry are increasingly engaged in implementing more sustainable and circular business models, and while it is recognized that some countries are championing supportive programmes, additional efforts are needed to create coherent policy frameworks which drive sustainability and circularity in the textile industry. Policies and legislative frameworks are required that enable businesses to shift to new business models without hindrance. Governments need to further regulate against toxic substances and harmful labour practices. Such enabling legislation needs to consider the whole industry and especially the hotspots, namely resource efficiency and chemicals in textile production, agricultural practices in natural fibre production and non-renewable resources use in synthetic fibre production, as well as addressing both social and environmental concerns.

Governments also have a role to play in enabling an inclusive and just transition and involving relevant stakeholders in the process, including those from affected communities and workforces as well as their representatives. There is a lack, especially, of implementation mechanisms to drive action. For example, encouraging sustainable resources consumption by incentives and tax rebates while discouraging the consumption of unsustainable resources by disincentives. The latter, especially, holds promise for stimulating demand for sustainable textiles, having the potential to use public procurement of textile products by local and national government agencies to pilot and promote new business models, such as selling services rather than products. The lack of capacity within governments to enforce legislation, and a lack of global coordination between governments, also need to be addressed if stronger governance is to be attained.

10.2 Collaboration and Financing

There is increasing recognition that leveraging existing solutions and best practices will not be enough to achieve a sustainable textile industry and that innovative solutions and new business models are required [47, 58]. However, there are gaps in technology, especially with regard to the systemic changes needed to move beyond small incremental improvements. There are also gaps in knowledge and experience with new business models, specifically in how to move away from existing business models to new circular and resource-efficient business models, and to provide the education, skills and support needed for new business models to flourish.

Significant support is thus needed for research and development into new business models and practices, and especially to accelerate the scaling of circular business models and sustainable solutions. It is required to bring public–private collaborations along with the involvement of all stakeholders including academia in order to strengthen the circular and sustainable solutions. Achieving the level and speed of change needed in the textile industry must involve all actors, especially the smaller manufacturers that have yet to make significant sustainability improvements. Unprecedented collaboration throughout the textile industry is required, creating a strong network of support, with extensive mentorship and capital investment. Such collaboration will instil the mindset that circularity is a value-chain-wide endeavour and that it needs to be embedded at the design stage.

Further, there is a lack of funding for developing and scaling the new business models. There is also a lack of funding for implementing more sustainable practices in the yarn, fabric and textile manufacturing stages of the industry, especially where these are small enterprises operating in developing countries. One important action is thus for partnerships to leverage funding from financial institutions, especially in those parts of the world where funding is difficult to leverage (and where the highest social and environmental impacts are occurring). There is a need for spaces and mechanisms which facilitate the deep level of collaboration required. In the context of the need for stronger governance and policies, governments in particular need such spaces for collaboration. With a considerable track record in projects directly involving textiles, as well as in the required associated disciplines (e.g. eco-innovation, life cycle thinking etc.) the United Nations is in a good position to provide such support, building on the strong base of existing networks and forums.

10.3 Change in Consumption Habits

Governments and brands/retailers are unlikely to take action at scale unless there is considerable advocacy. To this end, gaps in consumer awareness need to be addressed, and knowledge of and a preference for sustainable apparel and household textiles created among consumers. There is, especially, a need to address over consumption and fast fashion (acknowledging that in some parts of the world, clothing has to be affordable to meet basic human needs), as well as to instil consumption patterns in order to lessen the textile impact on environment and extend the life of garments. Furthermore, it also requires buy-in from consumers for new circular business models, such as clothing subscription-rental models, while reuse, repair/repurposing and recycling models require consumers to return their clothes to stores or collection depots and/or participate in sharing platforms/clothing exchanges. Educating and motivating consumers to play their role in the solution are critical.

This implies a sufficient number of forward-looking brands and retailers providing consumers with sustainable options so that consumers can exercise their purchasing power. New, innovative campaigns are needed that extend the reach of existing campaigns, for example using social media influencers and United Nations ambassadors to change lifestyle perceptions of what is "fashionable". Along with raising education and awareness, other options to motivate consumers should be implemented in order to incentivize sustainability practices by providing discounts. Most importantly, conditions must be put in place that makes it easy for consumers to choose sustainable options.

Furthermore, consumers need information if they are to be able to make ethical and sustainable choices. Consumer information tools, such as product labels, or trusted company-level analysis are thus important to enable better-informed decisions. However, product labels require better coordination to reduce confusion as well as actions to increase their applicability across products and improve their reliability and relevance (including better monitoring of social, economic and environmental effects of the textile industry to ensure the truthfulness of information). Brands/retailers and governments, working with civil society organizations, all have a role to play in implementing actions to change consumer behaviour.

11 Conclusions

The availability of resources, energy and water in a significant amount is essential for smooth industrial operations as well as for the textile industrial sustainability. Industries and other sectors can face unfavourable circumstances as the population will multiply and extreme seasonal variations, consequently, resulting in the availability of limited natural resources for the consumers. Incorporating technologies for the recycling and reuse of resources in the textile industry can result in sustainable consumption and production following the implications of nexus based circular economy such as the output of one process can be input for another process after being processed by certain technologies.

Agriculture, energy and water being the substantially consumed resources can optimize the textile industry's sustainability. In addition, the sustainability of the economy and the sustainability of people (workers and farmers) associated with this industry are also dependent upon the sustainability of textile industry; therefore, the challenge of sustainable textile production is required to be addressed. Furthermore, climate change is another challenge of concern that is not only affecting the agriculture production but also the depletion of natural resources. Apparently, climate change is affecting the textile industry as well. When it comes to the evaluation of resources consumed by the textiles, it has been analysed that cotton is 90% consumed in the textile industry and the remaining has been consumed by the people for managing their traditional quilts, pillows and other beddings. However, energy and water is a necessity for every sector not only limited to the textile industry. Another challenge is water supply which is a combination of groundwater and municipal water. As a result of excessive consumption of water for wet-processing, particularly the groundwater is getting lower and lower which can create a water crisis for the textile industry as well for the other water consumers in the city. Unfortunately, the wastewater after the production has been drained outside that is contaminating not only the soil around the factory premises but also the surface water. In order to reduce these ecosystem degrading impacts, the removal of toxins and dyes from the wastewater by installing the membrane and coagulation technologies cannot only conserve the environment but also influence the textile sustainability.

Daily textile production has been affected as a result of limited grid electricity and power outages in the country. Based on this study, it has been analysed that in order to meet the energy requirements of the factory, none of the textile factories in Faisalabad has installed any renewable energy system. Photovoltaic energy can be one of the effective options in renewables for this city, due to the fact that city has a prolonged summer season with intense solar irradiations. Another most convenient option of renewable energy for the textile factories is the utilization of environmentally exposed or burnt cotton crop waste as a source of biomass energy. Consequently, managing energy requirements adopting renewables for the factory and reducing the grid electricity load can be performed simultaneously. Government regulations and policies are so far lacking the influencing incentives and rebates for the installation of wastewater reclamation technologies and for the consumers of renewable energy technologies. Moreover, the extensive consumption of groundwater by the textile factories for wet-processing should be regulated in order to address these challenges for the textile sustainability. Nevertheless, the results of the study in this chapter based on the circular economy can be a supportive tool for the stakeholders of textile industry.

On the other hand, textile production is although impacting economic stability and industrialization in Pakistan. However, there are certain environmental and health risks associated with textile production as well, particularly, the burning of cotton crop leftovers, disposal of toxic wastewater into the external environment and the textile end-of-use. In spite of the fact, the provision of human services in the form of textiles and needed employment for the people, the industry carries social risks as well in the form of affecting the ecosystem and environmental health, and as a result, it will affect the human health. Although, cotton, energy and water are the common natural resources in the textile industry supply chain, however, the soil contamination with chemicals (sprays and fertilizers) for significant cotton production and dyes used in the printing is indirectly impacting human health. These health-hazardous activities are not only affecting the workers of the textile factories in particular but also the workers in the crop fields and wider communities where these activities are carried out.

The social and environmental effects from the textiles production are made harder to address because of business models that require speed and flexibility of production as well as manufacturing in locations where labour prices are lowest. The result is that textiles are predominantly manufactured in the cities where investment and employment are most needed, but where regulations protecting workers and the environment are weakest particularly in the cities such as Faisalabad, Multan and Sheikhupura. Despite the number of initiatives steadily improving the environmental and social performance of textiles, it is clear that more needs to be done. In particular, improvements need to move beyond incremental changes being made by large and high-end players to systemic changes undertaken by players of all sizes and market segments. Such systemic changes need to challenge the predominant business model of fast fashion, and rather than producing disposable items in bulk forms, produce recycled and repurposed textiles.

Moving towards sustainable and circular textiles demands the attention of all textile stakeholders from the supply sources to the workers, owners, policy makers and all the associated markets. A transformation from fossil fuels to renewables and efficient energy systems along with the alternative approaches of consuming toxic chemicals in the textile production. Recycling the clothes after their end-of-life consumption is another a challenge that needs to be addressed by involving innovative biodegradable textile materials. In order to benefit the industry, environment and society, new business models should be introduced and promoted established on the principles of circularity and sustainability. Achieving these changes will require coordinated actions by a range of stakeholders. Urgent actions should be taken towards the drivers of change that are the industrial policies, governance, collaborations and consumers engagement. A further overarching need is for real accountability across the supply and industry, whether this is achieved through specific transparency and traceability efforts or through collective programmes to drive systemic change.

Another way of sustainable and circular textile production can be supported by the UNEP. The objective of UNEP is to promote the major actions for the sustainability of textile industry such as providing leadership, knowledge development, partners' involvement, supporting alternate options of using toxins and transforming traditional textile production into a circular and sustainable production. This chapter has also identified the priority actions needed to advance circularity and sustainability in textile industry through an evidence-based approach.

The need of the hour is to convert ideas into actions by integrating academic, industry, research institutes, textile engineers and designers in order to bring a sustainable and circular change in the textile industry.

References

1. Böhringer C, Jochem PEP (2007) Measuring the immeasurable—a survey of sustainability indices. Ecol Econ 63(1):1–8. https://doi.org/10.1016/j.ecolecon.2007.03.008
2. Latif HH, Gopalakrishnan B, Nimbarte A, Currie K (2017) Sustainability index development for manufacturing industry. Sustain Energy Technol Assess 24:82–95. https://doi.org/10.1016/j.seta.2017.01.010
3. Bazilian M, Rogner H, Howells M, Hermann S, Arent D, Gielen D, Yumkella KK (2011) Considering the energy, water and food nexus: towards an integrated modelling approach. Energy Policy 39(12):7896–7906. https://doi.org/10.1016/j.enpol.2011.09.039
4. Hussien WEA, Memon FA, Savic DA (2017) An integrated model to evaluate water-energy-food nexus at a household scale. Environ Model Softw 93:366–380.https://doi.org/10.1016/j.envsoft.2017.03.034
5. Zamani B, Svanström M, Peters G, Rydberg T (2015) A carbon footprint of textile recycling: a case study in Sweden. J Ind Ecol 19(4):676–687. https://doi.org/10.1111/jiec.12208

6. Yang Y, Liu B, Wang P, Chen W-Q, Smith TM (2020) Toward sustainable climate change adaptation. J Ind Ecol 24(2):318–330. https://doi.org/10.1111/jiec.12984
7. PCI Wood Mackenzie (2016) Product developments in manmade fibres: is cotton able to compete? In: 33rd International cotton conference. Bremen
8. Quantis (2018) Measuring fashion: environmental impact of the global apparel and footwear industries
9. Albrecht TR, Crootof A, Scott CA (2018) The water-energy-food nexus: a systematic review of methods for nexus assessment. Environ Res Let 13(4):043002. https://doi.org/10.1088/1748-9326/aaa9c6
10. Kaddoura S, El Khatib S (2017) Review of water-energy-food Nexus tools to improve the Nexus modelling approach for integrated policy making. Environ Sci Policy 77:114–121.https://doi.org/10.1016/j.envsci.2017.07.007
11. Namany S, Al-Ansari T, Govindan R (2019) Sustainable energy, water and food nexus systems: a focused review of decision-making tools for efficient resource management and governance. J Clean Prod 225:610–626. https://doi.org/10.1016/j.jclepro.2019.03.304
12. Zhang S, Yi B-W, Worrell E, Wagner F, Crijns-Graus W, Purohit P, Varis O (2019) Integrated assessment of resource-energy-environment nexus in China's iron and steel industry. J Clean Prod 232:235–249. https://doi.org/10.1016/j.jclepro.2019.05.392
13. Brouwer F, Avgerinopoulos G, Fazekas D, Laspidou C, Mercure J-F, Pollitt H, Howells M (2018) Energy modelling and the Nexus concept. Energ Strat Rev 19:1–6. https://doi.org/10.1016/j.esr.2017.10.005
14. Silalertruksa T, Gheewala SH (2018) Land-water-energy nexus of sugarcane production in Thailand. J Clean Prod 182:521–528. https://doi.org/10.1016/j.jclepro.2018.02.085
15. Meng F, Liu G, Liang S, Su M, Yang Z (2019) Critical review of the energy-water-carbon nexus in cities. Energy 171:1017–1032. https://doi.org/10.1016/j.energy.2019.01.048
16. Nation T. (2018) Textile industry in Pakistan an open example of resistance economy. https://nation.com.pk/03-Jun-2018/textile-industry-in-pakistan-an-open-example-of-resistance-economy
17. Wadho W, Chaudhry A (2018) Innovation and firm performance in developing countries: the case of Pakistani textile and apparel manufacturers. Res Policy 47(7):1283–1294. https://doi.org/10.1016/j.respol.2018.04.007
18. Guarnieri P, Trojan F (2019) Decision making on supplier selection based on social, ethical, and environmental criteria: a study in the textile industry. Resour Conserv Recycl 141:347–361. https://doi.org/10.1016/j.resconrec.2018.10.023
19. Baloch MH, Tahir Chauhdary S, Ishak D, Kaloi GS, Nadeem MH, Wattoo WA, Hamid HT (2019) Hybrid energy sources status of Pakistan: an optimal technical proposal to solve the power crises issues. Energ Strat Rev 24:132–153. https://doi.org/10.1016/j.esr.2019.02.001
20. Imran M, Özçatalbaş O, Bashir MK (2018) Estimation of energy efficiency and greenhouse gas emission of cotton crop in South Punjab. Pakistan. J Saudi Soc Agri Sci. https://doi.org/10.1016/j.jssas.2018.09.007
21. Government of Pakistan, Economic Survey 2018–2019
22. De Oliveira Neto GC, Ferreira Correia JM, Silva PC, de Oliveira Sanches AG, Lucato WC (2019) Cleaner production in the textile industry and its relationship to sustainable development goals. J Clean Prod 228:1514–1525.https://doi.org/10.1016/j.jclepro.2019.04.334
23. Barma MC, Saidur R, Rahman SMA, Allouhi A, Akash BA, Sait SM (2017) A review on boilers energy use, energy savings and emissions reductions. Renew Sustain Energy Rev 79:970–983. https://doi.org/10.1016/j.rser.2017.05.187
24. Dilaver M, Hocaoğlu SM, Soydemir G, Dursun M, Keskinler B, Koyuncu İ, Ağtaş M (2018) Hot wastewater recovery by using ceramic membrane ultrafiltration and its reusability in textile industry. J Clean Prod 171:220–233. https://doi.org/10.1016/j.jclepro.2017.10.015
25. Sandin G, Peters GM (2018) Environmental impact of textile reuse and recycling—a review. J Clean Prod 184:353–365. https://doi.org/10.1016/j.jclepro.2018.02.266
26. Khan S, Yufeng L, Ahmad A (2009) Analysing complex behaviour of hydrological systems through a system dynamics approach. Environ Model Softw 24(12):1363–1372. https://doi.org/10.1016/j.envsoft.2007.06.006

27. Kojiri T, Hori T, Nakatsuka J, Chong T-S (2008) World continental modeling for water resources using system dynamics. Phys Chem Earth Parts A/B/C 33(5):304–311. https://doi.org/10.1016/j.pce.2008.02.005
28. Mereu S, Sušnik J, Trabucco A, Daccache A, Vamvakeridou-Lyroudia L, Renoldi S, Assimacopoulos D (2016) Operational resilience of reservoirs to climate change, agricultural demand, and tourism: a case study from Sardinia. Sci Total Environ 543:1028–1038
29. Qi C, Chang N-B (2011) System dynamics modeling for municipal water demand estimation in an urban region under uncertain economic impacts. J Environ Manage 92(6):1628–1641. https://doi.org/10.1016/j.jenvman.2011.01.020
30. Stave KA (2003) A system dynamics model to facilitate public understanding of water management options in Las Vegas Nevada. J Environ Manage 67(4):303–313. https://doi.org/10.1016/S0301-4797(02)00205-0
31. Simonovic SP (2002) World water dynamics: global modeling of water resources. J Environ Manage 66(3):249–267. https://doi.org/10.1006/jema.2002.0585
32. Dai J, Wu S, Han G, Weinberg J, Xie X, Wu X, Yang Q (2018) Water-energy nexus: a review of methods and tools for macro-assessment. Appl Energy 210:393–408. https://doi.org/10.1016/j.apenergy.2017.08.243
33. Henry B, Laitala K, Klepp IG (2019) Microfibres from apparel and home textiles: prospects for including microplastics in environmental sustainability assessment. Sci Total Environ 652:483–494
34. Almroth C, Bethanie LÅ, Roslund S, Petersson H, Johansson M, Persson N-K (2018) Quantifying shedding of synthetic fibers from textiles; a source of microplastics released into the environment. Environ Sci Pollut Res 25(2):1191–1199
35. De Falco F, Gullo MP, Gentile G, Di Pace E, Cocca M, Gelabert L, Brouta-Agnésa M, Rovira A, Escudero R, Villalba R, Mossotti R, Montarsolo A, Gavignano S, Tonin C, Avella M (2018) Evaluation of microplastic release caused by textile washing processes of synthetic fabrics. Environ Pollut 236:916–925
36. Zambrano M, Pawlak J, Daystar J, Ankeny M, Cheng J, Venditti R (2019) Microfibers generated from the laundering of cotton, rayon and polyester based fabrics and their aquatic biodegradation. Mar Pollut Bull 142:394–407
37. Roos S, Jönsson C, Posner S, Arvidsson R, Svanström M (2019) An inventory framework for inclusion of textile chemicals in life cycle assessment. Int J Life Cycle Assess 24(5):838–847
38. Xu X, Hou Q, Xue Y, Jian Y, Wang L (2018) Pollution characteristics and fate of microfibers in the wastewater from textile dyeing wastewater treatment plant. Water Sci Technol 78(10):2046–2054
39. Barnes D, Galgani F, Thompson R, Barlaz M (2009) accumulation and fragmentation of plastic debris in global environments. Philos Trans Royal Soc London B: Biol Sci 364(1526):1985–1998
40. Barrows A, Cathey S, Petersen CW (2018) Marine environment microfiber contamination: global Patterns and the diversity of microparticle origins. Environ Pollut 237:275–284
41. Stanton T, Johnson M, Nathanail P, MacNaughtan W, Gomes R (2019) Freshwater and airborne textile fibre populations are dominated by natural Not Microplastic, Fibres. Sci Total Environ 666:377–389
42. Machado DS, Abel A, Kloas W, Zarfl C, Hempel S, Rillig MC (2018) Microplastics as an emerging threat to terrestrial ecosystems. Glob Change Biol 24(4):1405–1416
43. Kirstein I, Kirmizi S, Wichels A, Garin-Fernandez A, Erler R, Löder M, Gerdts G (2016) Dangerous hitchhikers? Evidence for potentially pathogenic vibrio spp. on microplastic particles. Mar Environ Res 120:1–8
44. Newbold T, Hudson LN, Arnell AP, Contu S, De Palma A, Ferrier S, Hill SLL, Hoskins AJ, Lysenko I, Phillips HRP, Burton VJ, Chng CWT, Emerson S, D Gao, Pask-Hale G, Hutton J, Jung M, Sanchez-Ortiz K, Simmons BI, Whitmee S, Zhang H, Scharlemann JPW, Purvis A. (2016) Has land use pushed terrestrial biodiversity beyond the planetary boundary? A global assessment. Science 353(6296):288–291

45. Turley DB, Copeland JE, Horne M, Blackburn RS, Stott E, Laybourn SR, Harwood J (2009) The role and business case for existing and emerging fibers in sustainable clothing: final report to the department for environment, food and rural affairs (Defra)
46. Canopy (2018) The hot button report 2018: ranking of viscose producer performance. Time to move on next generation solutions
47. Ellen MacArthur Foundation (2017) A new textiles economy: redesigning fashion's future
48. WWF (2007) Cleaner, greener cotton: impacts and better management practices
49. UNEP (2019) Global chemicals outlook II. United Nations environment programme. UNEP circularity platform
50. UNEP (2013) Global chemicals outlook: towards sound management of chemicals
51. Kant R (2012) Textile dyeing industry an environmental hazard. Nat Sci 04(01):22–26
52. KEMI (2014) Chemicals in textiles. Risks to human health and the environment
53. Greenpeace (2018) Destination zero: seven years of detoxing the clothing industry
54. Rex D, Okcabol S, Roos S (2019) Possible sustainable fibers on the market and their technical properties. Fiber bible part 1
55. Watson D, Nielsen R, Palm D, Brix L, Amstrup M, Syversen F (2016) Exports of nordic used textiles: fate, benefits and impacts. Nordisk Ministerråd, Copenhagen
56. Leal Filho W, Ellams D, Han S, Tyler D, Paço A, Moora H, Balogun AL (2019) A review of the socio–economic advantages of textile recycling. 218:10–20
57. Wetengere KK (2018) Is the banning of importation of second-hand clothes and shoes a panacea to industrialization in East Africa? 6(1):119–41
58. GFA and BCG (2018) Pulse of the fashion industry
59. Halog A, Anieke S (2021) A review of circular economy studies in developed countries and its potential adoption in developing countries. Circ Econ Sustain. https://doi.org/10.1007/s43615-021-00017-0
60. GIZ (2019) Circular economy in the textile sector
61. Shahbaz, Abbas Lin Han, Chiang Hsieh Kuaanan, Techato Juntakan, Taweekun (2020) Sustainable production using a resource–energy–water nexus for the Pakistani textile industry. Journal of Cleaner Production. 271122633-10.1016/j.jclepro.2020.122633

Circular Economy in Product Development—A Case Study

M. Gopalakrishnan, R. Prema, and D. Saravanan

Abstract Textiles and apparels are fundamental parts of the everyday life. Due to the awareness on global warming, every stakeholder is working on the waste utilization and reducing the stress on newer resources. New business model, circular economy, deals with the effective use of the waste materials and generated at various stages of the product life cycle. The circular economy concept is applied in all the fields of engineering including the textiles and apparels. The production of textiles and apparels involves huge amounts of human work and leaves substantial amounts of carbon footprint. This chapter explains three difference cases, involving, recycling of wastes to products with better values, upcycling for different end uses.

Keywords Carbon footprint · Circular economy · Fibre waste · PET bottles · Reuse · Spinning waste · Waste utilization

1 Introduction

Cotton fibres are the most well-known staple fibres, and the length of the strands runs somewhere in the range of 22 and 32 mm, which develop as single cells from the cotton seed. Each cotton seed can have 5000 to 20,000 fibres including linters, shorter fibres of 1.5–10 mm long, held fast and grounded to the seeds [1]. After harvesting, the strands are removed from the seed and utilized for yarn production. Recycling of wastes generated at different stages plays an important role in textile industry. Many

M. Gopalakrishnan (✉)
Department of Textile Technology, Bannari Amman Institute of Technology, Sathyamangalam 638401, India

R. Prema
Department of Electrical and Electronics Engineering, Bannari Amman Institute of Technology, Sathyamangalam 638401, India

D. Saravanan
Department of Textile Technology, Kumaraguru College of Technology, Coimbatore 641 049, India

S. S. Muthu (ed.), *Circular Economy*, Environmental Footprints and Eco-design of Products and Processes, https://doi.org/10.1007/978-981-16-3698-1_4

interventions and research works have been done on the recycled natural and man-made fibres, and few studies have also been conducted by researchers to convert the recycled fibres into useful products. There are few patents in this line, which mainly deal with conversion to different useful end products.

Nowadays, the disposal of waste is a biggest issue and the prevention of waste is one method to reduce the waste. The reuse of waste into a useful products can reduce the impact of waste in a great attitude. The 3R's, reuse, reduce and recycle can reduce the waste disposal [2]. Today, major brands are aiming to use the recycled fibres in their products with the label of recycled fibres and it has very good response in consumers. Another important parameter is the waste developed during the yarn production. These waste is not suitable to produce good quality yarns. However, it can be used to produce some other useful products. The life cycle assessment (LCA) brought you the environmental impacts of a particular products. The grocery bag produced with recycled fibres has very low LCA than compare with the bag produced in virgin fibre [3].

The purpose of shopping bag is distinguished with the number of times it is used and the type of material [4]. If the shopping bag is used for single use, then the environmental impact will be more. On the other hand, if it is used more than one time, then the environmental impact is less than compared with single use. Therefore, the life of the shopping bag is increased for reduction of environmental impacts. Muthu is a well-known researcher in this field, and he made biggest impact in evaluating the environmental impacts for the shopping bags [5–7]. Muthu also assessed the LCA for the production of various textile fibres [3].

The author also defined the eco-functional assessment, a new concept, much ahead of LCA. The author applied eco-functional assessment on the shopping bag to analyse the impact on recycled fibres [8–10].

Brownstein reveals a nonwoven textile-based air filter produced from polyester fibres produced using recycled polyethylene terephthalate (PET) beverage bottles [11]. Yuan also explains the technique for producing nonwoven using recycled polyester waste materials with intrinsic viscosity of greater than 0.7 [12]. Heifetz disclosed a method for manufacturing the yarns using recycled cotton wastes, [13] and a process for using recycled waste cotton materials to produce a textile product is provided by Lightman using three different categories of waste cotton materials that are obtained from textile formation processes [14]. No studies have been reported to reuse the rotor waste into a useful products like grocery bags. In this case studies, the production of nonwoven shopping bag is from the waste of rotor.

In the modern world, the main problem faced by the society is plastic wastes, especially plastic bags, which take minimum 500 years to degrade in the soil, thereby contributing to climatic change and damage to the environment in large scale. Plastic bags are expensive and hard to clean or remove from the environment. Plastic bags are not easy to recycle, and the actual recycling rate for plastic is around 1%. Nonwoven polypropylene (PP) bags are produced as an alternate for other plastic bags, but they are also harmful to the environment because they also come under the plastic category. They are produced by converting the polypropylene fibre into sheet form and finally joined by spun bonding method. These polypropylene bags are also non-degradable but instead of degrading, and they are broken down into individual fibres, as an effort to recycling [15]. Many governments have banned the plastic bags and polypropylene bag usages in many applications, which has led to emerging of paper bags, woven bags as the alternate for the plastic and PP bags though the cost of the bags is relatively higher when compared to plastic bags [16].

The understanding of the circular economy and its practical applications to economic systems evolved incorporating different features and contributions the idea of closed loops. The concept of circular economy within the textiles industry often refers to the practice of clothes and fibres being recycled, to re-enter the economy as many times and as much as possible rather than ending up as wastes or landfills. In a circular economy practice, typically raw materials are extracted, manufactured into differential commercial goods, bought, used and eventually discarded by consumers in a cyclic-way. This chapter presents three case studies, where upcycling has been used to develop value-added products from the industry wastes generated at different stages.

2 Case Study 1: Upcycling of Cotton Wastes

Large amounts of cotton fibres are thrown as the wastes by the rotor spinning mills and garments manufacturing units during cutting process. It has been reported that the total amount of wastages generated in India annually is about 80,000–85,000 tonnes [17]. These wastes are recycled into fibres using opening (willowing) machines [18]. Interestingly, these recycled cotton fibre wastes are either used for bedding process and further dumped in the soil or burnt in the open environment. The present invention (case study) reduces the wastages of recycled coloured cotton fibres (Fig. 1) and convert them into useful products through upcycling.

Fig. 1 Recycled cotton wastes—rotor spinning wastes

Recycled coloured cotton fibre wastes and natural resin like (gelatin, starch and PVA) are used as the raw materials to produce nonwoven sheets that can be converted into bags. Recycled coloured cotton fibres are converted into nonwoven fabrics by needle-punching and further bonded intimately using chemical bonding method. Chemical bonding is one of the methods to produce nonwoven fabrics by using synthetic or natural resins. According to the research, many products are produced using recycled cotton fibre wastes but no work has ever been done related to the production of nonwoven carry bag from recycled coloured cotton fibre wastes using gelatin, starch and polyvinyl alcohol as a binding agent.

In garments industry, during cutting of knitted cotton fabrics, large amount fabrics are thrown as wastes. Recycled coloured cotton fibre wastes (depending on the colour of the waste materials) are converted into web form using carding machine. The waste knitted cotton fabrics are recycled into fibres using recycling machines that consist of 5–7 doffers, which help in tearing the fabrics and convert them into fibres. Development of prototype was carried out through miniature-carding machine consists of three main parts like licker in, cylinder and doffer. Each sample consists of 50 g of recycled coloured cotton fibre wastes. The speed of the miniature carding was maintained at 2 m/min, i.e. the output speed of the web. By increasing or decreasing the speed of the carding machine, the thickness of the web was varied and the web consists of multiple layers of the fibres. Figure 2 shows the web formed in miniature -carding machine using recycled fibres.

The web produced from recycled coloured cotton fibres does not possess adequate strength due to short length of the fibres. So, fine yarns, with the linear density of 80–100 s, are inserted between the webs to reinforce and increase the strength with the ratio of yarn to web 10:1, i.e. 10 yarns are inserted in 1 m of web, with the distance between each yarn is maintained at 10 cm. The webs are separated into two layers, and the yarns are inserted between the two layers of web. After inserting the yarns, the two layers are combined together and moved to the next process. The reinforced yarns (4) are inserted in between the web (5). For inserting the reinforcing yarns, the webs are separated into two parts (2, 3) as shown in Fig. 3.

The webs created from the recycled coloured cotton fibres do not have enough strength to convert them into nonwoven fabrics. So, binding agents are used to create bond between the fibres to increase the strength of the web. Three binding agents were used, in different combinations, in this invention to create bond between the fibres, gelatin, starch and polyvinyl alcohol. These binding agents are selected due to its

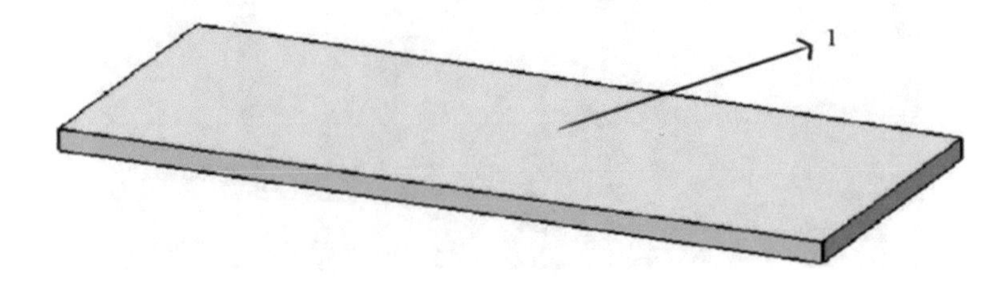

Fig. 2 Web from recycled coloured cotton fibre wastes

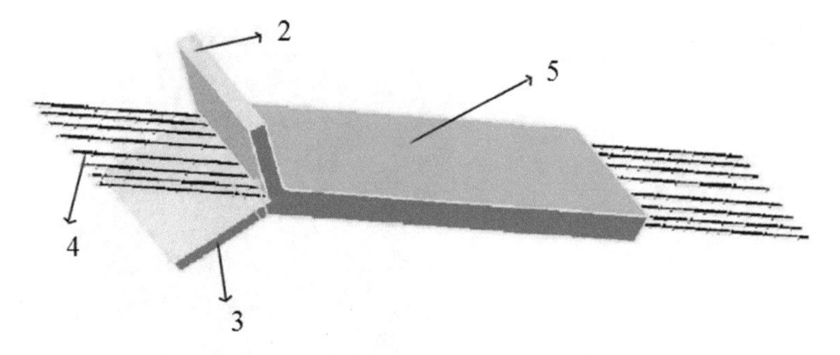

Fig. 3 Insertion of reinforcing yarn between the webs

eco-friendly nature and cost efficiency. Figure 4 shows the binding agents that were applied on the web spray method. The spray unit (6) is placed just above the fabric layer, and the binding agent (7) is applied/sprayed onto the web at high pressure. The web which was applied with the binding agents may have the excess amount of binding agents. This should be removed from the web by passing the web through a high-pressured drum (pressure will be maintained according to the thickness of the web). Due to the pressure on the web, the excess amount of binding agents is be squeezed out and collected separately.

Figure 5 shows the removal of the excess binding agents using calendar pressure rollers (8, 9). The binding agents applied web (10) was passed between the pressure rollers. Due to the nip pressure, the excess amount of binding agents is squeezed out from the web, and the finished web (11) is taken and cured. The pressure roller spreads the binding agents evenly to all the fibres by applying pressure. The binding agents are fixed on the fibres by passing the webs through hot oven, at 100–120°C. The nonwoven fabrics are then collected in a roller after the drying process. According to the required size, the nonwoven fabrics are cut into the desired shapes and sizes using a cutting machine. The nonwoven fabrics are spread in the lay form over the spread table. The numbers of lays depend on the quantity of carry bags produced.

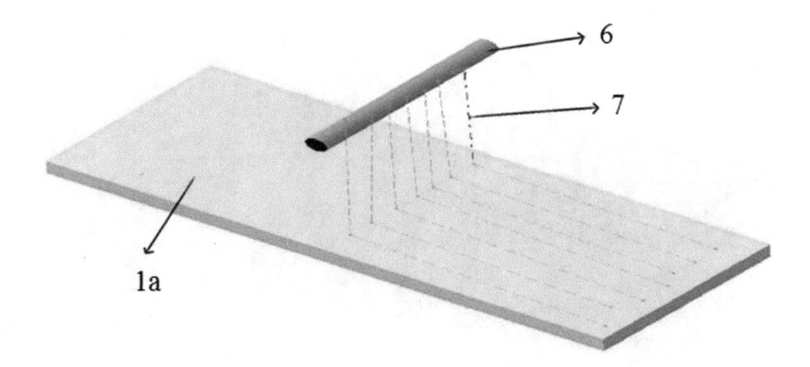

Fig. 4 Application of binding agents on web

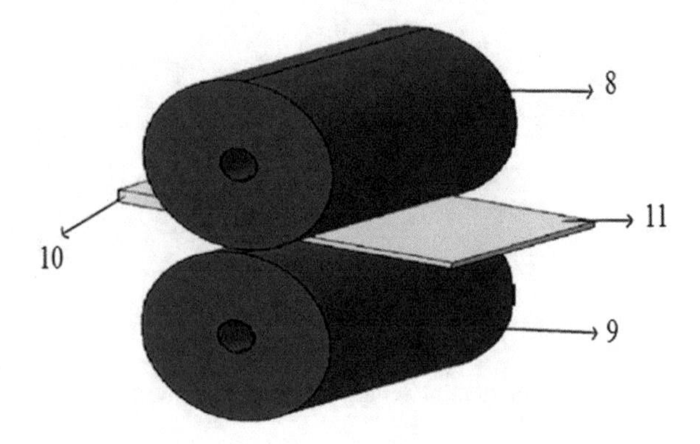

Fig. 5 Calendar roller for padding

Straight knife cutting machine, band knife cutting machine and hand shears are used to cut the nonwoven fabrics.

The cut fabric pieces are separated into different parts like front part, back part and side panels. Reactive dyes are used to print the fabrics, and mainly front and back parts are printed. Different designs are printed on the fabrics. After the printing process, fabrics are dried by hot oven to fix the reactive dye to the fabric, inside the hot oven for 5–10 min under 80–100°C. After the printing process, the printed and cut fabrics are joined together by stitching process. In the present invention, the over-lock stitch is used to join the side parts of the carry bag to increase the strength of the carry bags. On the other hand, straight stitches are used to stitch the straight line in the handle part of the carry bag. Figure 6 shows the process of stitching. The front pattern (12), side (14) and back (13) were joined to together to make the bag. The lining material was used to cover the raw edges of the bag. Finally, the handles (15) were attached to body of the bag. The handle of the carry bags is also made of cotton yarns. After stitching the carry bags, the handles are fixed by stitching process. There are different handles used for different carry bags and that depends on the size and style of the carry bags. There is another method used to fix the handles in the carry bags without using stitching process. After converting into web form, yarns are

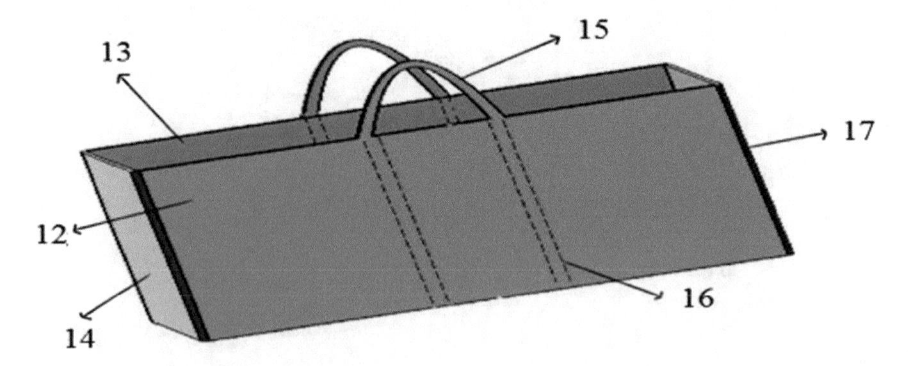

Fig. 6 Recycled cotton fibre nonwoven bag

inserted between the two webs to increase the strength of the web. While inserting the yarn between the webs, the handles are also inserted along with the yarns. This method eliminates the stitching process of handle and also increases the strength of the carry bags.

3 Case Study 2—Recycling of PET

Polyester is a man-made fibre that is made from petroleum-based raw materials, a non-renewable resource. The usage of polyester has increased several times since its discovery though it is also non-biodegradable. Among the total global fibre production, 49% is polyester fibre and more than 63,000 million tonnes of polyester fibre are produced, annually. The clear PET water bottles are used as the raw material for production of recycled polyester. The polyester waste can be recycled either mechanically or chemically. The first PET bottle was recycled in the year 1977. Recycling of plastic bottles can prevent them from going into landfills that cause pollution [19, 20].

In chemical recycling, the waste plastic is made into its original monomer and then the fibres are produced. Chemical recycling is not successful due to its high cost and the need for large quantities to achieve the process economy [21]. In mechanical recycling, the plastic bottle is washed, shredded and made into polyester chips. The fibres are then made using those chips. The process of recycling of PET bottles in mechanical method is shown in Fig. 7.

Fig. 7 Mechanical recycling of PET bottles

Collection, sorting and flaking of PET bottles: The used PET bottles are collected from the consumers or wastes, compacted, baled and transported to the flake production unit [22]. The bottles are sorted according to the requirement (colour/quality). The clear bottles are sorted for the production of white polyester yarn, and the green bottles are used for the production of green polyester yarns [23]. Automatic sorting is done by colour identification method [24]. The labels in the bottles are removed and cleaned. The bottles are chopped into flakes by the shredder with rotating blades to 0.4–8 mm. These flakes are then allowed to float in water to separate high density materials from the flakes. The flakes may be washed using detergent and 2% NaOH with hot water and then rinsed with cold water or using tetrachloroethylene. Then, the flakes are dried at a temperature of 17 °C for nearly 6 h [25] to a level less than 0.02%.

Production of fibres: The flakes are directly melted and extruded to produce the fibres (flake—fibres) or converted into chips/pellets and processed (flake—pellets—fibre). The melt extrusion method is used to produce fibres [22] at a temperature greater than 240°C. The filaments can be cut into staple fibres. The recycled polyester fibres can be used separately to produce fabrics or can be blended with other fibres to acquire a required property from the fabrics. Some of the strong, durables items like coat, jacket, bags, shoes and accessories are produced from these fibres. Sleeping bags, carpets, fibrefill, carboot linings are some of the recycled polyester fibre product [23]. The biggest difference between rPET and PET is its impact on the environment. The requirement of energy and water for the production recycled polyester is lesser then that required for the production of virgin polyester. The energy consumption for the production of recycled polyester is only 2/3 of the energy required for the production of virgin polyester. Recycled polyester requires only 10% of water used for production of virgin polyester.

3.1 Advantages of Recycled Polyester

- **Raw material**—Recycled polyester reduces the need for extraction of crude oil and natural gas from earth. The pollution caused during the extraction of virgin polyester can also be reduced [26].
- **Lesser energy and water consumption**—The energy consumption for the production of recycled polyester is only 2/3 of the energy required for the production of virgin polyester.
- **Reduces pollution**—The plastic bottles if not recycled go into waste as landfills. Landfills cause land and soil pollution which harms some of the living creatures. Recycling of polyester emits lesser hazardous gas in the environment than the gas released during the extraction of virgin polyester [26].
- **Closed-loop system**—The garments can be produced from recycled polyester without any degradation in quality. The garment produced from recycled polyester can be recycled again. Thus, the cycle continues and forms a closed-loop system [27].

The abrasion and wear testing of fabric give the data to predict the materials and its coating durability. The abrasion tester (Martindale) was used to test abrasion resistance of the fabrics. The fabric to be tested is cut using the template. Four samples are cut and are clamped onto the lower plate (sample holder) of the abrasion tester. Sand paper is used as abrade in this test. A load of 400 g is applied, and the number of rotation cycles was set as 200. The specimen holders move in a complex motion. After the specified number of cycles, the samples were weighed and the abrasion resistance is calculated using the given formula.

Abrasion resistance (%) = (Mass after abrasion/Mass before abrasion) $\times$ 100

Abrasion resistance of virgin and recycled polyester is listed in Table1. The results show that the samples exhibit (virgin polyester is 93% and recycled polyester is 95%) almost equal abrasion resistance. In fact, the recycled polyester exhibits slightly better Abrasion resistance than that of virgin polyester. This may be the nonlinear structure and cross linking of polyester in recycling.

The multi-directional resistance to rupture of a fabric specimen was evaluated by bursting strength. After the specimen is clamped, the pressure builds up on the glycerin in the cylinder underneath the fabric. The resulting pressure is conveyed to the rubber diaphragm. At certain pressure the fabric bursts. Then, the bursting strength was noted and the results are tabulated in Table 2.

Bursting strength of virgin and recycled polyester was listed in Table 2. And it shows that the bursting strength of the virgin polyester is 114 lbs, and the recycled

Table 1 Abrasion resistance of polyester

S.No	Fabric type	Specimen weight (g)		Weight loss	Abrasion resistance
		Before abrasion	After abrasion		
1	Virgin polyester	0.244	0.226	0.018	92.62%
		0.237	0.221	0.016	93.25%
		0.234	0.218	0.016	93.16%
		0.237	0.220	0.017	92.83%
2	Recycled polyester	0.259	0.248	0.011	95.75
		0.248	0.232	0.016	93.55%
		0.252	0.241	0.011	95.63%
		0.249	0.237	0.012	95.18%

Table 2 Bursting strength

S.No	Bursting strength in lbs	
	Virgin polyester	Recycled polyester
1	114.5	78.31
2	113.7	79.88
3	115.6	79.5
Mean	114.6	79.23

polyester is 79 lbs. This indicates that the virgin polyester possesses good bursting strength than that of recycled polyester. The reduction in strength may be the reduced crystallinity during recycling. Even the recycled polyester is bursting strength than that of the other textile fibres, and it may suitable to spin the fibre. The bending length of the fabric was calculated as per the standard procedure. The result of bending length is tabulated in Table 3.

The bending length of the virgin polyester is 1.7 cm, and the recycled polyester is 1.25 cm. The bending length of the virgin polyester is greater than recycled polyester. This may be the stiffness and crystallinity of virgin polyester is better than recycled polyester (Table 4).

The wicking property of recycled polyester is better when compared to that of virgin polyester. The reduction in crystalline of recycled polyester may increase the wicking behaviour.

The washing fastness of recycled polyester and virgin polyester was done as per the ISO test methods. The results were displayed in Table 5. The results indicate that the both recycled and virgin polyester showed good wash fastness. So, the recycling doesn't influence the dyeing behaviour of the fabric. Overall, the recycled polyester

Table 3 Bending length of the polyester fabric

S.No	Virgin polyester		Recycled polyester	
	GSM	Bending length	GSM	Bending length
1	1.332	1.6	1.360	1.2
2	1.332	1.9	1.360	1.3
3	1.332	1.7	1.360	1.2
Mean	1.332	1.7	1.360	1.4

Table 4 Wicking test

S. No	Virgin polyester		Recycled polyester	
	Time (seconds)	Distance travelled by water (cm)	Time (seconds)	Distance travelled by water (cm)
1	30	0.2	30	2.0
2	60	1.0	60	3.3
3	120	1.4	120	3.9
4	300	2.0	300	5.3
5	600	2.9	600	6.5

Table 5 Wash fastness

Sample no	Fabric type	Grade
1	Virgin polyester	Very good
2	Recycled polyester	Very good

fibre exhibits good in almost all the testing except the strength. So, the recycled fibre made from PET bottles is an alternative to the virgin polyester.

4 Case Study 3—Recycled Fibre-Based Mulching Sheets

Mulch is a covering material produced using either organic or inorganic substances and applied on the soil surface, in cultivation of various crops. Mulching is carried out for varieties of reasons including soil moisture retention, reducing soil evaporation, heat trapping, reducing run-off losses, increasing germination, weed prevention and control, protection of roots from temperature fluctuations and control the soil erosion [28–30]. Soil mulching may contribute to closing the yield gap between attainable and actual yields. Usage of plastic mulch increases the yield of the crops by 18% and 27%, respectively, for wheat and maize [31]. However, there has been an increase in the greenhouse gas emissions (yield-scaled) while using plastic mulching sheets to the extent of 32 and 10% for these crops. It has been observed [32] that use of plastic mulch in the ridge-furrow plots results in higher crop yield in clay and silt loam soils and both straw mulch and plastic mulch significantly increase maize yields and water utilization efficiency under different humidity and growing-season precipitation, compared to control plots (without mulching).

Majority of the mulching films are made up of PE (a resistant synthetic polymer) which causes a serious environmental drawback consisting of a huge quantity of wastes. Different coloured PE mulches reflect the light rays differently and thereby influence the conditions of cultivation and yield of the crops. Plastic mulches are widely used, and black polyethylene has been an attraction in many situations due to its low cost and the proven results in crop production. Soil temperature is always less in the case of white mulch compared to that of red and black coloured mulch sheets [33]. Mulching film residues are found to be as high as 15.3% of the total quantity of films used, depending upon the film thickness, mulching time and crop type and the highest number of residues are found in the cotton fields [34]. Using black polypropylene nonwoven mulch shows a positive effect on soil temperature and amount of water required during growth stage of potato crops with reduced biomass of weeds by almost 89% [35]. However, use of plastic films is limited by the financial cost but also by the cost of the collection and recycling of the plastic residues [30]. Some of the biodegradable mulches do not meet the expectations of the farmers and often degrade too quickly or incompletely at the time of harvesting and thereby necessitating the manual removal of mulch sheets [36]. One advantage of paper mulches is that easy disposal of residual material in the soil itself [37].

Starch-based biodegradable plastic mulch films exhibit strong negative effect on growth of wheat plant during both vegetative and reproductive growth both above-ground and below-ground parts. Nevertheless, presence of earthworms has overall positive effect on growth and eliminates the negative impacts of plastic residues [38]. Biodegradability of mulching films comprising of starch and polybutylene adipate terephthalate under aerobic conditions has been studied. Straw mulching is limited by the availability of straw in the field, which is often being used also for feeding ruminants or as biofuel [30]. Textile-based mulching material has been compared with conventional polyethylene mulch, mechanical mowing and chemical controls that are widely used in citrus orchards [39]. Cotton recycled wastes, carding wastes and comber noil wastes (Fig. 8) are converted into web through need punching and sandwiched between lightweight polypropylene nonwoven web to produce mulching sheet, suitable for growing greens and other vegetables. Greens cultivated using such mulching sheet show better growth in terms of number of leaves for a given number of days allowed for growth after sprouting (Fig. 9).

Fig. 8 Cotton Wastes **i** open ends spinning wastes **ii** comber noil wastes

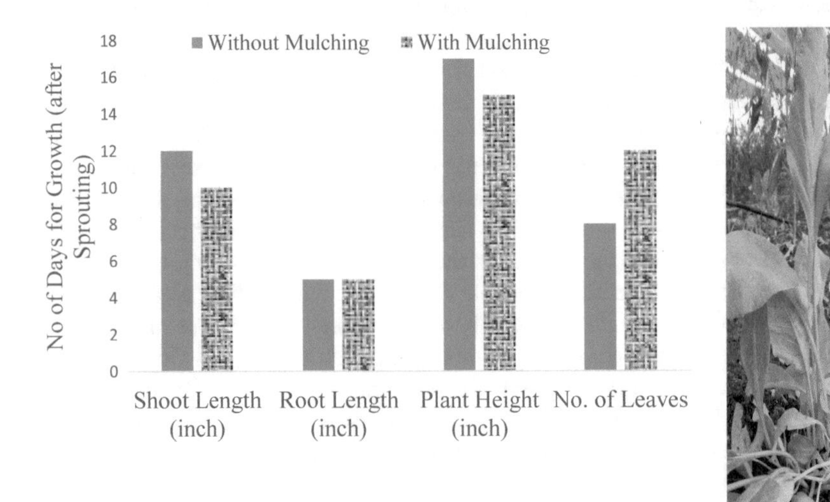

Fig. 9 Comparison of Plant Growth **i** with and without Mulching Sheet **ii** Plant

5 Conclusion

Today, the people are more aware of sustainability and global warming. The recycled materials are the ones that reduce the global warming and increase the sustainability. In this chapter, different case studies have been presented based on the wastes obtained from fabrics, in-process wastes obtained from spinning and post-consumer product to provide a comprehensive understanding to the readers.

References

1. Palme A (2017) Recycling of cotton textiles : characterization, pretreatment, and purification. Dissertation, Chalmers University of Technology
2. Muthu SS (2015) Measuring the reusability of textile products. In: Muthu SS (ed) Handbook of Life Cycle Assessment (LCA) of textiles and clothing. Elsevier, Netherland, pp 83–92
3. Muthu SS, Li Y, Hu JY et al (2012) Eco-impact of plastic and paper shopping bags. J Eng Fiber Fabr 7:26–37. https://doi.org/10.1177/155892501200700103
4. Muthu SS (2015) LCA of cotton shopping bags. In: Muthu SS (ed) Handbook of Life Cycle Assessment (LCA) of textiles and clothing. Elsevier, Netherland, pp 283–299
5. Muthu SS, Li Y, Hu JY, Mok PY (2012) Quantification of environmental impact and ecological sustainability for textile fibres. Ecol Indic 13:66–74. https://doi.org/10.1016/j.ecolind.2011.05.008
6. Muthu SS, Li Y, Hu J-Y, Mok P-Y (2012) Recyclability Potential Index (RPI): The concept and quantification of RPI for textile fibres. Ecol Indic 18:58–62. https://doi.org/10.1016/j.ecolind.2011.10.003
7. Muthu SS, Li Y, Hu JY, Mok PY (2012) Eco-functional assessment combined with life cycle analysis: concept and applications. Energy Educ Sci Technol Part A Energy Sci Res 29:435–450
8. Muthu SS, Li Y, Hu JY et al (2013) Assessment of eco-functional properties of shopping bags. Int J Cloth Sci Technol 25:208–225. https://doi.org/10.1108/09556221311300228
9. Muthu SS, Li Y, Hu J-Y et al (2013) Modelling and quantification of eco-functional index: the concept and applications of eco-functional assessment. Ecol Indic 26:33–43. https://doi.org/10.1016/j.ecolind.2012.10.018
10. Muthu SS, Li Y, Hu JY, Mok PY (2011) Carbon footprint of shopping (grocery) bags in China, Hong Kong and India. Atmos Environ 45:469–475. https://doi.org/10.1016/j.atmosenv.2010.09.054
11. Brownstein JM, Brownstein K (2014) High efficiency low pressure drop synthetic fibre based air filter made completely from post consumer waste materials. US Patent 8398752B2, 04 August 2010
12. Yuan CF (2008) The technique for producing nonwoven cloth by using recycled polyester waste material with intrinsic viscosity of greater than 0.7 or more parts, China Patent No. 101173420, 2008
13. Heifetz SD (1994) A method for manufacturing the yarns using recycled cotton wastes. US Patent 5,331,801, 26 Jul 1994
14. Lightman ED (2011) A textile product can be provided that includes yarn that can have at least three different categories of waste cotton material that are from textile formation processes. US Patent 20110250425, 12 Apr 2011
15. Thompson RC, Moore CJ, vom Saal FS, Swan SH (2009) Plastics, the environment and human health: current consensus and future trends. Philos Trans R Soc B Biol Sci 364:2153–2166. https://doi.org/10.1098/rstb.2009.0053

16. Dash DK (2019) Packaged water industry warns govt on plastic ban. Times of India, https://timesofindia.indiatimes.com/india/classifying-packaged-water-bottles-as-single-use-plastic-will-impact-rs-30000-crore-industry/articleshow/71054974.cms. Accessed 12 Feb 2021
17. Wanassi B, Azzouz B, Ben HM (2015) Recycling of post-industrial cotton wastes: quality and rotor spinning of reclaimed fibers. Int J Adv Res 3:94–103
18. Halimi MT, Ben HM, Azzouz B, Sakli F (2007) Effect of cotton waste and spinning parameters on rotor yarn quality. J Text Inst 98:437–442. https://doi.org/10.1080/00405000701547649
19. Paszun D, Spychaj T (1997) Chemical recycling of poly(ethylene terephthalate). Ind Eng Chem Res 36:1373–1383. https://doi.org/10.1021/ie960563c
20. Gopalakrishnan M, Subrata Das, Akshaya S, et al (2020) Preparation and processing of recycled polyester for environmental sustainability in apparel industry. Colourage March:29–36
21. Welle F (2011) Twenty years of PET bottle to bottle recycling—an overview. Resour Conserv Recycl 55:865–875. https://doi.org/10.1016/j.resconrec.2011.04.009
22. Shen L, Worrell E, Patel MK (2010) Open-loop recycling: a LCA case study of PET bottle-to-fibre recycling. Resour Conserv Recycl 55:34–52. https://doi.org/10.1016/j.resconrec.2010.06.014
23. Eric Joo, Jee-Eun oh (2019) Challenges facing recycled polyester. Text World July:1–4
24. Shen L, Worrell E (2014) Plastic recycling. In: Emst Worrell and Markus A. Reuter (ed) Handbook of recycling. Elsevier, Netherland, pp 179–190
25. Webb H, Arnott J, Crawford R, Ivanova E (2012) Plastic degradation and its environmental implications with special reference to poly(ethylene terephthalate). Polymers (Basel) 5:1–18. https://doi.org/10.3390/polym5010001
26. van Elven M (2018) How sustainable is recycled polyester? Fash United. Available via DIALOG. https://fashionunited.uk/news/fashion/how-sustainable-is-recycled-polyester/2018111540000. Accessed 15 Mar 2021
27. Textile Exchange (2016) Recycled Polyester commitment. https://textileexchange.org/recycled-polyester-commitment/?gclid=Cj0KCQjw0emHBhC1ARIsAL1QGNeLVPfI7X7-VUsRklcpavnYpyyE9oXDl0p1R36lWHMlAmEIms0rCe4aArqVEALw_wcB. Accessed 15 Mar 2021
28. Shaw DA, Pittenger DR, McMaster M (2005) Water retention and evaporative properties of landscape mulches. In: Proceedings of annual irrigation show. Nov:134–144.
29. Telkar SG, Singh AK, Kant K et al (2017) Types of mulching and their uses for dry-land condition. Biomol Reports 9:1–4
30. Qin W, Hu C, Oenema O (2015) Soil mulching significantly enhances yields and water and nitrogen use efficiencies of maize and wheat: a meta-analysis. Sci Rep 5:1–13. https://doi.org/10.1038/srep16210
31. He G, Wang Z, Li S, Malhi SS (2018) Plastic mulch: tradeoffs between productivity and greenhouse gas emissions. J Clean Prod 172:1311–1318. https://doi.org/10.1016/j.jclepro.2017.10.269
32. Yu Y-Y, Turner NC, Gong Y-H et al (2018) Benefits and limitations to straw- and plastic-film mulch on maize yield and water use efficiency: a meta-analysis across hydrothermal gradients. Eur J Agron 99:138–147. https://doi.org/10.1016/j.eja.2018.07.005
33. Fortnum BA, Decoteau DR, Kasperbauer MJ (1997) Colored mulches affect yield of fresh-market tomato infected with meloidogyne incognita. J Nematol 29:538–546
34. Zhang D, Liu H, Hu W et al (2016) The status and distribution characteristics of residual mulching film in Xinjiang, China. J Integr Agric 15:2639–2646. https://doi.org/10.1016/S2095-3119(15)61240-0
35. Dvořák P, Hajšlová J, Hamouz K et al (2009) Black polypropylene mulch textile in organic agriculture. Lucr Ştiinţifice Ser Agron 52:2–6
36. Hannan JM (2012) Black degradable plastic mulch evaluation ISRF11–36:25–26.
37. Saravanan D (2019) Changing facets of mulching materials. Curr Trends Fash Technol Text Eng 5:61–62. https://doi.org/10.19080/CTFTTE.2019.05.555659

38. Qi Y, Yang X, Pelaez AM et al (2018) Macro- and micro- plastics in soil-plant system: effects of plastic mulch film residues on wheat (Triticum aestivum) growth. Sci Total Environ 645:1048–1056. https://doi.org/10.1016/j.scitotenv.2018.07.229
39. Kitiş YE, Kolören O, Uygur FN (2017) Yeni Tesis Mandalina Bahçesinde Malç Tekstili Uygulamasının Yabancı Ot Kontrolü ve Mandalina Gelişimine Etkileri. Turkish J Agric—Food Sci Technol 5:568. https://doi.org/10.24925/turjaf.v5i6.568-580.729

Circular Economy Implementation in Chilean Retail Industry

Sebastian Garcia Jarpa, Anthony Halog, and Lorna Guerrero

Abstract Chile imports an estimate of 75% of the textile products in its market to which Asian countries provide up to the 70% (Monroy in Viste Consciente, Chile University, Architecture and Urbanism Faculty, 2020 [1]). These Asian goods are brought mostly by biggest multinational retail corporations in Chile: Falabella, Paris, and Ripley, covering 62.1% of the entire textile market in the country (Monroy in Viste Consciente, Chile University, Architecture and Urbanism Faculty, 2020 [1]). Due to the large amount of textile products available in the market, there is a dramatic growth of discarded goods that only in Chile represent an estimated of 550 tons per year of textile products that end up in landfills (La Tercera in Ecocitex: Eliminar el desecho textil en Chile, 2021 [2]) as result of the lack of recycling programs that translate into strong environmental problems (Ministerio del Medio Ambiente, Chile. Hoja de Ruta Nacional a la Economía Circular para un Chile sin Basura, 2020 [3]). The REP Law was created regarding garbage management. It aims to reduce the amount of garbage going to landfills with specific goals for 2030 and 2040 in an attempt of the government to support Circular Economy development in Chile. However, textile company disposal is not considered on this law. Despite of that, some retail companies in Chile are implementing long term Circular Economy projects considering environmental, economic, and social programs targeting at becoming sustainable markets. On the production side, these companies have implemented their own sustainable lines which include local textile manufacturers and commercialization of products made by local communities, highlighting the artisanal work and traditional technics with alpaca, organic cotton, and other recycled material. On the reduction side, they are collaborating with independent recycling businesses that take care of the reduction of discarded textile in Chile which are run under Circular Economy models. Recycling companies such as Rembre, Ecocitex, Travieso, Ecofibra, and Retex have

S. G. Jarpa (✉) · L. Guerrero
Universidad Técnica Federico Santa María, Valparaíso, Chile

L. Guerrero
e-mail: lorna.guerrero@usm.cl

A. Halog
Queensland University, Brisbane, Australia
e-mail: a.halog@uq.edu.au

S. S. Muthu (ed.), *Circular Economy*, Environmental Footprints and Eco-design of Products and Processes, https://doi.org/10.1007/978-981-16-3698-1_5

seen in disposed textile a new profitable market opportunity and an environment focused course of action towards reducing the negative impact of one of the biggest contaminating markets in the world. Nevertheless, the lack of traceability regarding the volume of textile available in the market makes it extremely difficult to create a roadmap of products needed to implement a Circular Economy screening processes to quantify the volume of textile reaching landfills and the magnitude of their impact on the reduction of this waste. With all this in consideration and based on sustainability reports from retail and recycling companies, a Circular Economy value chain map has been created as a description of textile products follow in Chile.

Keywords Circular economy · Textile · Chile · Retail · Sustainability development goals · Recycling

1 Introduction

In Chile, most of the textile products sold in local markets are imported. The vast majority of those products come from China and other Asian countries. The suppliers from the eastern countries offer low prices and have established international trading agreements with Chile that are part of the neoliberal model of the latest [4] (Fig. 1).

The trading records among Chile and Asian countries are not recent. The national textile production was weakened since the start of the neoliberal system in 1980, as it welcomed strong competitor enter the national markets. The Chilean textile manufacturing cut off their productions and international wholesale brands rose up to take over the textile industry, affecting the economic models of the national textile culture. This new textile cultural model and the opening of its economy to the world have transformed Chile into the most textile consuming country in the region, with an average of 50 products purchased per year [1].

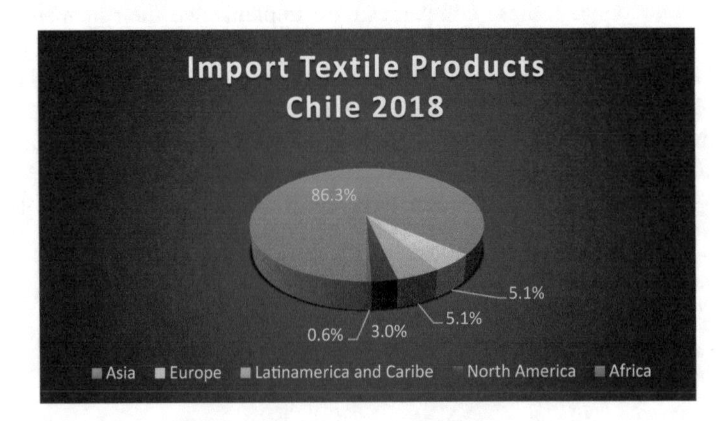

Fig. 1 Import textile products in Chile, 2018 [5]

The textile market is controlled mostly by 3 important retail companies: Falabella, Paris, and Ripley, which ad up to 62.1% of the total textile market in Chile [2]. They compete against international retail corporation such as H&M and Zara due to the fast fashion model where they keep ever changing trends in one-week seasons meaning a high amount of products at low cost [2].

Due to this situation mentioned above, there is a high volume of textile products that end in landfills. Different actors have committed to transforming the textile value chain. Companies such as Ecocitex, Retex, and Ecofibra were created with the mission of implementing actions in order to reduce the accumulation of discarded textile products. These companies, referred to as "recycling companies" in this chapter, have an important role in trying to close the loop of the textile products value chain in Chile, as they are achieving agreements with retail companies. Their participation in the value chain is increasingly important to relief the pressure in landfills due to the rising amount of materials not reaching them.

The Chilean Government is implementing an ambitious strategy to follow a circular economy model in all their industrial sectors and processes. The plan is focused mainly on increasing recycling rates, as an attempt to decrease the pressure of big waste volumes in landfills, since studies have shown that their useful life is 12 years if no improvement is made [3]. However, textiles are not included on this program, a major setback to transitioning from linear to circular model.

This research is focused on the most important retail companies in Chile: Falabella, Paris, and Ripley; and their attempts to follow a sustainable system with the aim to create a map of textile market in Chile, with special focus in the textile value chain in Chile.

Firstly, an in-depth analysis of sustainable programs of these three important retail companies is done. The advancement of environmental awareness, social development, labour conditions and textile value chains to develop a sustainable system is presented, with the purpose of showing how these companies implement circular economic ideas in the business and how far they are from a 100% sustainable development in their markets.

Secondly, the development of new recycling companies related to textile industry is shown with the aim of building circular models with disposed textile products and how its development can create a new textile value chain of discarded textile products from retail companies, creating alliances to treat textiles that are disposed in landfills in the previous years. Additionally, the vision of every recycling company and the relationship of circular economy models is presented.

Finally, a textile value chain map is created based on data collected, presenting connections between the sources of textiles products and the country, the business model, and the destinations of its value chain are presented, illustrating how the development of textile industry in Chile is evolving based on three aspects: import products, retail companies, and recycling companies.

Conclusions include an analysis of the textile industry in Chile, presenting the most important companies involved and how recycling companies have treated discarded textiles, avoiding big volumes of textile products from going to landfills in the country.

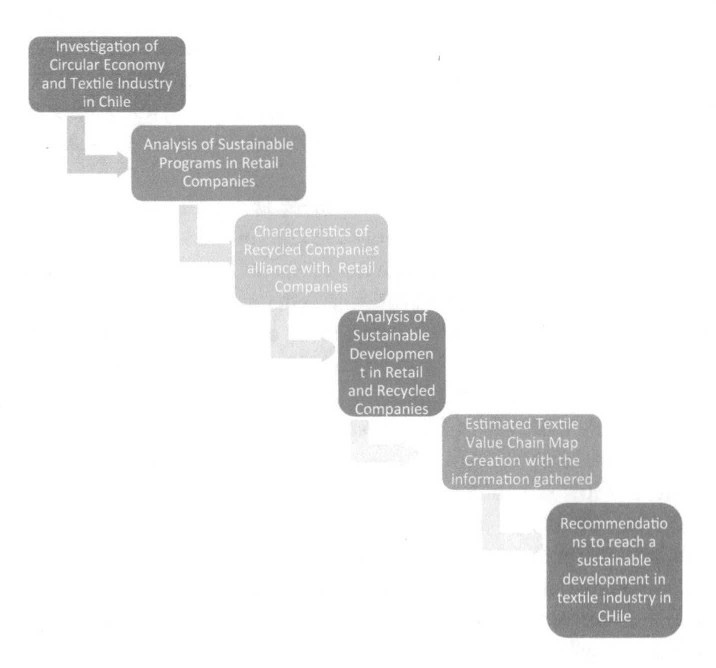

Fig. 2 Roadmap of the research about sustainable textile industry in Chile

Recommendations present ideas on how the value chain could improve in the future and challenges of the textile industry in Chile to reach a 100% sustainable system.

The information is gathered from sustainability reports from retail companies, and articles related to the increase of recycling companies dedicated to the textile industry. Unfortunately, there was not enough scientific information about textile industry in Chile in previous years, not allowing to create a complete and quantifiable map of value chain of textile industry. However, this work will be the first step towards an accountable analysis of textile market, as well as an overview of the progression in the transformation to circular economy model (Fig. 2).

2 Textile Products Cycle in Chile

The textile products come from Asian countries are imported at a low price. The most important textile market are the retail companies, which it is a mix of retail companies, international brands, and local entrepreneurs. Retail companies are associated with recycling companies to treat their waste textile volumes and give them other uses with the transformation of clothes into different products: bags, denim, personal accessories, as examples.

2.1 *Retail Companies*

The most important retail textile markets have started to improve their processes with the aim to be more sustainable in the future. Their initiatives have started in the previous years (around 2016), discussing about how important to consider a circular economy model in their businesses [7–9].

They have strong programs to improve the working conditions of their employees, implementing different training programs related to the inclusion of migrants, the gender equality, leadership of women and others.

2.2 *Recycling Companies*

Recycling companies are a key step to increase the percentage of recycling rate of textile products in Chile. These companies are mostly created by entrepreneurs with a motivation to help the country to be more sustainable, recovering waste generation of textile products and transforming into new sustainable products to incorporate to the market. An important consequence of the creation of this kind of companies is the reduction of textile products that ends in landfills, decreasing the percentage of volume that is accumulated on these sectors.

Recycling companies are linked with big retail companies to help them in the recycling program of their businesses, receiving certain amount of textile products volume that is variable during a year. These alliances have permitted to help each other: retail companies to reach their goals to have more awareness with the environment, and recycling companies to coexist in the market with a constant raw material for their businesses.

A big change has started in Paris company, for example. They started to send waste textile products to an external company located in Germany, I:CO for their proper treatment. However, in 2019, the company has started an alliance with Rembre, a Chilean company responsible to receive and transforming the textile products material into new products. When a national company is chosen to treat waste volume the cost will be less, due to the distance from the source.

In terms of fair conditions of workers, retail companies have implemented several activities and new ideas to give more value to their workers, which they have offered different leadership program trainings.

2.3 *Social and Energy Organizations*

Energy companies are connected with other companies by having sustainable actions, but not directly with textile products. The use of sustainable electricity consumption with higher percentage of renewable sources for textile stores is an important action that must be maintained in the future. Chile has a big potential of renewable sources and this advantage has been used by textile markets. In this way, Ripley has an ambitious plan to use only renewable sources in a short-term period. Falabella, on the other hand, has reached 40% of renewable sources for their markets and it is estimated that Falabella wants the same as Paris, reaching 100% of sustainable energy sources.

Clothes donations to organizations like Coaniquem, Abriendo puntos, and Maria Ayuda are some examples of the destination of clothes in good conditions that is disposed by retail companies. The connection of these organizations is directly with recycling companies, which has two objectives: the reuse of textile products in good conditions, and a social objective to help people with needs.

The next will be an analysis of important retail companies in Chile in order to know deeply every company and their actions to be more sustainable in the future, complementing the information presented above.

3 Retail Companies in Chile: Falabella Case Study [7]

Falabella is one of the biggest retail companies in Chile with 44 stores, 31,000 workers and 16,000 suppliers with a Directory composed by 22% women and 78% men. Its development in sustainability is wide ranging including projects and advances in all aspects: social, environmental, and technological issues.

The interest of Falabella is to be a leader in sustainable model. To follow this plan, the company has created a material theme matrix based on interviews and surveys with suppliers, community, clients, and chiefs from the company. This matrix describes the pillars of sustainability, escalating the importance of interest of the company versus the business. On this way, Falabella has 4 principal strategies to include in their businesses: Integrity; Significative Experience; More Social Value; and Climate Action and Circularity. Every point and the mostimportant actions will be described in the next sections.

3.1 Integrity

Integrity involves two sustainable goal developments: Goal 16 (Peace, Justice, and Strong Institutions) and Goal 17 (Partnerships for the goals). The training programs have reached 76% of its workers in 2019, which includes 18.14 h average of training, audited by BH Compliance, a company dedicated to certifying and monitoring sustainable developments. This program has permitted to standardize their workers in terms of what is important for a sustainable growth of the company, involving their workers and giving them more importance in the retail company. The integrity is applied in the company with important trainings for workers: about politics of Falabella, crime prevention and E-learning and their advantages for workers (Fig. 3).

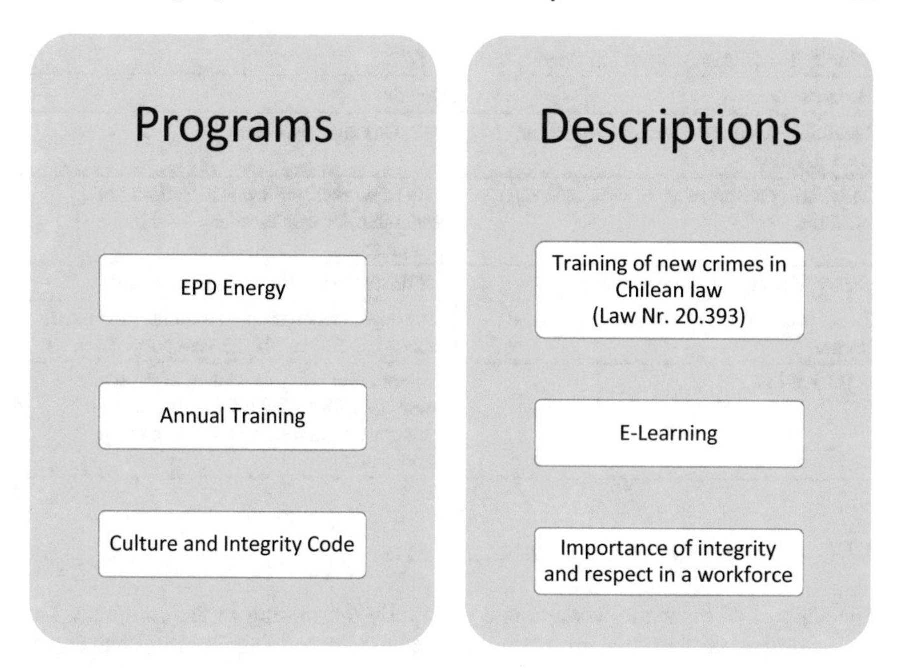

Fig. 3 Training programs for workers of Falabella in 2019 [9]

3.2 *Significative Experience*

Sustainable goals are divided in development of Clients and Workers. Experiences are related to a better experience for clients and workers of Falabella are the next ones: Goal 3 (Good Health and Well-being), Goal 4 (Quality and Education), Goal 5 (Gender Equality), Goal 8 (Decent Work and Economic Growth), Goal 10 (Reduced Inequalities), Goal 12 (Responsible consumption and production (12). A significative experience is focused on clients and workers in Falabella company.

3.2.1 Clients Better Experience

Technology developments were included in Falabella to deliver a better experience for clients and workers, a step further into a globalization and a connection of important advances in retail companies. For example, the creation of Falabella App has allowed to reach more clients in Chile (1.6 million of clients use the app), giving them a more personalized service, reducing the time of transports to stores and to facilitate the sale process. The first automated Delivery Centre of Falabella in Chile called "Click and Collect" is a pilot project in order to improve the experience of clients and to give a faster service of sale.

Table 1 Benefits for women in Falabella company [7]

Benefits	Details
Flexibility to complement the labour and personal life	Fair working conditions
Monitoring the presence of women in high positions	64% of workers are women. Training in leadership for women at any level in the company
Salary with equity management	1.8 times the minimum salary in Chile
Benefits to maternity and paternity on the norms	Free days, kindergarten, masculine post-natal, money per child as the most important benefits
F Generation	Promotion of talented women in the retail company with digital training that last 6 months, organized by a social start-up, Laboratoria

3.2.2 Workers Better Experience

The support of Falabella workers is a constantly developing in the company. The opportunity of professional growth and formation with training programs and evaluations with feedbacks are important actions of the company to improve the workforce. These kinds of evaluations allow an effective communication and integration in the activities of every position. The performance evaluation is a tool to have access to professional development programs, and it is divided in three parts: Corporative Evaluation, Performance Retail Evaluation, and Labour Satisfaction.

Falabella has an objective in 2020 to reinforce the gender equality at any level of hierarchy in the company, where two third parts of the workers are women, and implementing them training of leadership in the company with mentoring programs and internal networking between women under the company. The training is complemented with a series of benefits for workers, highlighting the respect and awareness about their workers (see Table 1).

Additionally, Falabella counts with programs to integrate young talented professionals to the organization and scholarships, having links with important universities in the country to develop projects of young professional interests. This involvement of young professionals is according to Pacific Alliance of Young Employability since 2017. There are two programs to highlight in this theme, described as follows (Table 2).

Workers also have other important benefits in the company and an important role for the company, which 44% of them are part of unions, and 93% of workers has collective agreements related to safe labour place, health benefits, and equilibrium between personal and professional life.

Salaries of workers are regulated by the International Position Evaluation, which the company is worried about having a higher salary than the minimum wage in Chile, with 1.8 points more than the minimum salary in the country.

Table 2 Programs for young talented professionals in Falabella Company [7]

Program	Description
Hackaton	Attracting the best digital talented in collaboration with Laboratoria organization, offering formation of 6 months
Talent bank	Offering exchange programs in coordination with Talentum company to learn about good practices are responsible to give new tools and skills to their positions in the company. In 2019 for example, there were 83 workers participating, with 38 workers promoted

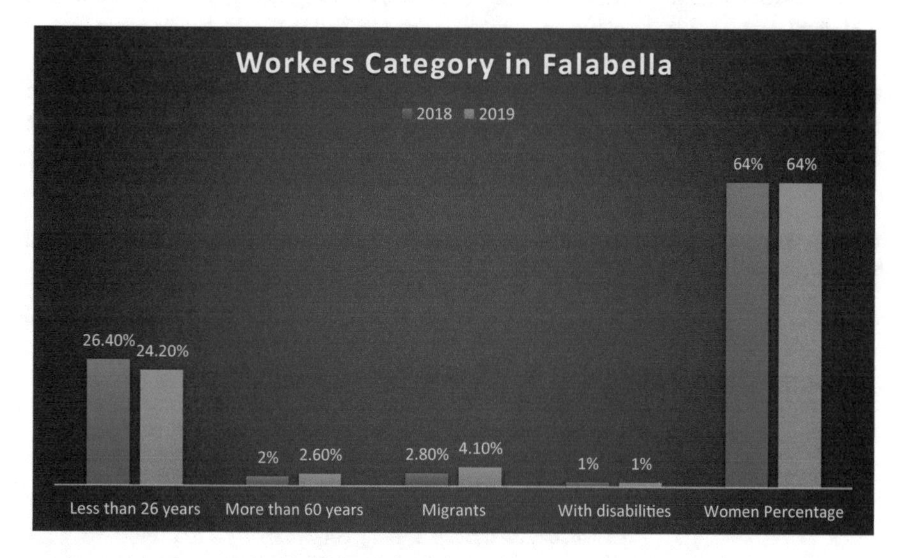

Fig. 4 Workers category in Falabella (2018-2019) [7]

The happiness and the comfort of their workers is important in Falabella. Their principles of the organization describe a promotion of safe labour environment, healthy and in equilibrium. Stores are certified by OHSAS 18,001 (Health and Safety Management) and with joint committees to offer a comfortable and ergonomic place to work (Fig. 4).

3.3 *More Social Value*

Falabella has 17.000 suppliers and 1500 sellers, which 4890 were new in 2019. In this way, Falabella supports the development of independent vendors since 2017. The Sustainable Goals pursued in this section are: No Poverty (#1), Good Health and Well–Being (#3), Quality Education (#4), Decent Work and Economic Growth (#8), Industry, Innovation, and Infrastructure (#9), and Reduced Inequalities (#10).

To manage the transportation of products, Falabella has a traceability program to align suppliers of owned brands to improve their social and environmental practices. The traceability is done with two conditions (Table 3).

Results of audit in 2019 are shown in Table 4, a way to manage the textile products they sell in Chile, a step more to reach a sustainable model.

Falabella has a growth of local entrepreneurship in their products, having 90% of national vendors. The compromise with local entrepreneurship is also important to get value of the community. Falabella has a priority of payment for independent local vendors, making them less than 30 d, and offering training for communities that need.

For local entrepreneurship, Falabella have prepared training programs to develop and incorporate new tools to increase their businesses. For example, Brand Fashion Market is an event for emergent designers to promote their products. E-Commerce training is other platform to train local entrepreneurship in their businesses.

Table 3 Traceability program of Falabella with their suppliers to implement a sustainable social and environmental model in their business [7]

Conditions	Description
Vendor agreement (VA)	Operational and commercial conditions related to payment deadlines, delivery dispatches, quality, working conditions, etc.
Vendor compliance (VC)	Ethical and Responsible standards with social audits under SMETA protocol registered in SEDEX program, a database of vendors and suppliers of the company. Parameters evaluated are human rights, management system, free choice employment, right to association, health and security, child labour, discrimination, regular employment and subcontracts, force working conditions, right to work, environmental management, business ethics, and benefits for the community

Table 4 Audit results with national and Asian suppliers [6]

	Audit results (89%)	Audit results (89% accredited)
	National workshops (%)	Asian workshops (%)
Immediate action	1	0
High risk	5	32
Medium risk	25	46
Low risk	26	21
Acceptable	43	1

3.4 Climate Action and Circularity

Sustainable Goals Development related to Climate Action and Circularity in Falabella are: Affordable and Clean Energy (#7); Industry, Innovation, and Infrastructure (#9); Climate Action (#13).

One of actions is the strict quality control inspection implemented for their products from different suppliers, with the purpose of reducing waste and managing the quality and compliance of sustainable certifications. Part of the quality control is the application of international alarm on product reparations, diagnosis of the cycle value chain related to the quantity of waste and optimization programs, and the compliance of quality certifications from their suppliers.

Energy reduction is included in the sustainable plan of Falabella, using resources in an efficient way to minimize the environmental impact. There has been an important consumption electrical reduction where 40% comes from renewable energy. This reduction was supported using LEED technology, the control of circuits and optimization of climate equipment. With this action, Falabella has received the LEED certification: The Leadership in Energy and Environmental Design of their stores.

CO_2 emissions of Falabella have been considered to analyse and see what the most important factors of CO_2 emissions generation are. The study has found that a high percentage is done by textile product transportation (92%), followed by workers transportation (6%), and waste generation (2%) as showed in (Figs. 5 and 6).

Falabella has been working hard on reducing its carbon footprint, reaching the level 3 Greenhouse Protocol (GHG Protocol) in 2019, for all installations in Chile. With this level, Falabella has been awarded with the recognition of "Huella Chile"

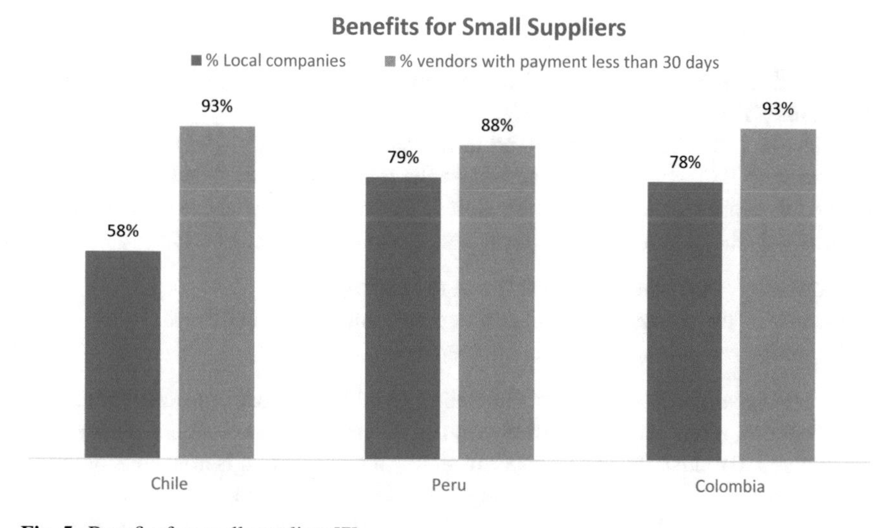

Fig. 5 Benefits for small suppliers [7]

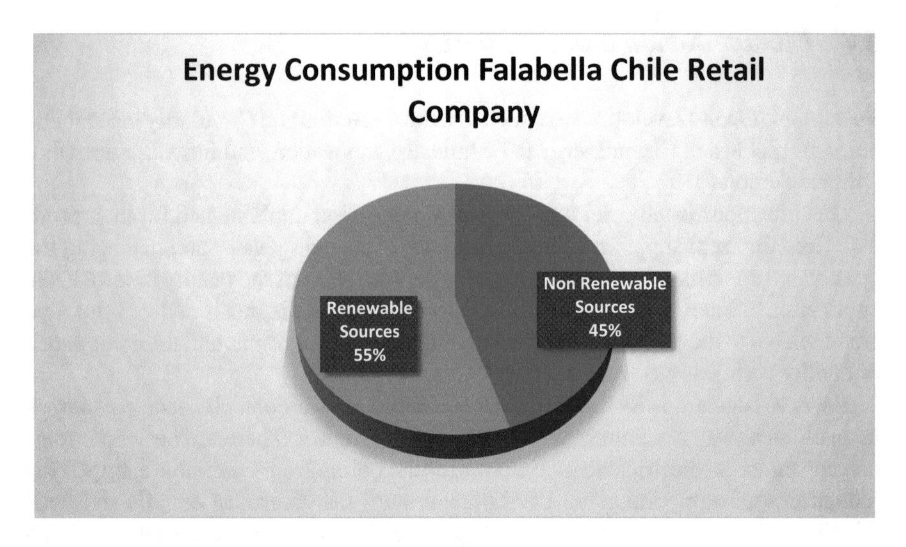

Fig. 6 Energy consumption Falabella Chie Retail Company [7]

Table 5 Carbon Footprint Measurement, measured as CO_2 equivalent [6]

Measurement	CO_2 equivalent
Carbon footprint Chile 2019	112.219 [ton] CO_2 equivalent
Level 1	14.680 [ton] CO_2 equivalent
Level 2 (market)	18.721 [ton] CO_2 equivalent
Level 2 (location)	41.851 [ton] CO_2 equivalent
Level 3	78.817 [ton] CO_2 equivalent

Program from Environmental Ministry of Chile, award the effort of greenhouse gases emissions (GHG) (Table 5 and Fig.7).

Falabella has other actions related to the packaging area of their products. The alliance with FSC about good practices in the management of forest resources has allowed to avoid plastic bags in their stores. These plastic bags have changed into a more sustainable material: paper bags. The agreement FSC has permitted.

- Promoting good practices in the forest management.
- Increasing the presence of FSC products in commercial businesses.
- Migrating operational elements into certified products.

Following with the use of paper, Falabella has reduced the use of ticket of purchase, implementing a technological ticket sent to personal emails of clients. This action has allowed to reduce the use of paper in stores, a material that is now considered as unnecessary to deliver a payment, using technological solutions.

Falabella has a green category called *"Categoria + Verde"* which it is platform with sustainable textile products. Production from recycled materials, reused or with

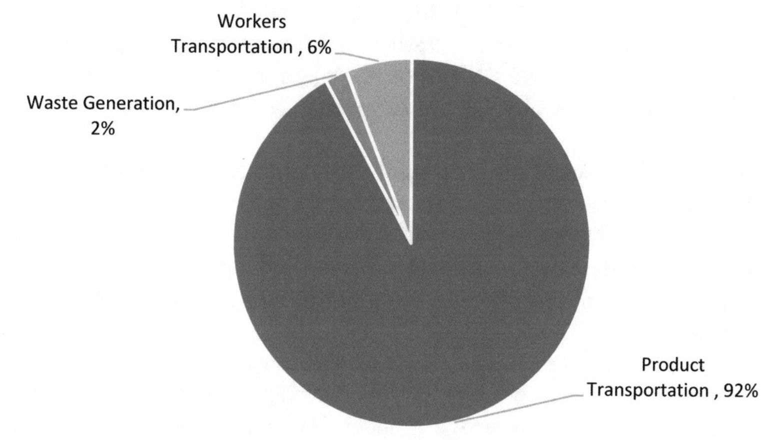

Fig. 7 CO_2 emissions distribution in Falabella retail company [6]

Table 6 Sustainable actions of the + Verde Category of Falabella [7]

Action	Description
Reuse	Helps to generate less waste and to reduce disposable elements
Conscious materials	More conscious raw materials with the environment
Recycling	Reused raw materials or recycled: upcycling and circular economy
Community	Products that offer value to the work of social risk people, with projects of innovation and creativity

a low impact in the environment are promoted on this section of the company, with a sustainable vision with important 4 actions (Table 6).

In " + Verde" section of Falabella, the increase of green products has been important during previous years with 368.7% of more products offered in stores. The number of brands has reached 52 at the end of 2019. The politics of Falabella is still working on including more sustainable brands to offer wide ranging options in the future for their clients (Table 7).

Falabella has developed its own sustainable brand of products based on recycled raw materials recovered and treated by REMBRE, a 'B company' (compromised by social and environmental impacts) which recycles retail textiles from Falabella

Table 7 Green Products and Brands in Falabella company [7]

	July 2019	December 2019
Green products numbers	403	1.889
Green category brands	41	52

stores and are sorted out by type of material. These materials are transformed into new products like pillow filling, Denim products, and eco-bags. Also, Falabella is making an effort to change PVC textiles into Polyethylene textiles and reuse this in the making of other products like trash bags, films, and scotch tape. This project translates into an approximate 4.8 tons of polyethylene cloths of high density from other recycled sources. Finally, the overall process avoids the production of 6.3 ton of CO_2 emission.

Also, among their sustainability projects, the company is currently carrying out the removal of one-use plastic hangers and bags. Up to 2019, it was able to spare the introduction of 30.5 tons of new plastic and cut down 15% of hanger by reusing them. Regarding plastic bags, Falabella has prevented the use of 500,000 bags per year until 2019. This means 344 tons of polyethylene per year, which represents a budget reduction of around 5 to 7%.

To continue with a sustainable waste management, Falabella has presented a project where they commit to the country by reaching a zero-waste production and keeping accountable for their textile products. With these two recycling actions, Falabella has recycled 60% of their waste generated (Figs. 8 and 9)

In Conclusion, Falabella has a strategy to implement sustainability mostly in the client experience, the relationship with employees, work environment, and recycling processes, in order to implement a full circular economy model in the future, which means commit to the reduction of the volume of discarded textile, improving their relationship with their workers, and a better sale experience for clients (Table 8).

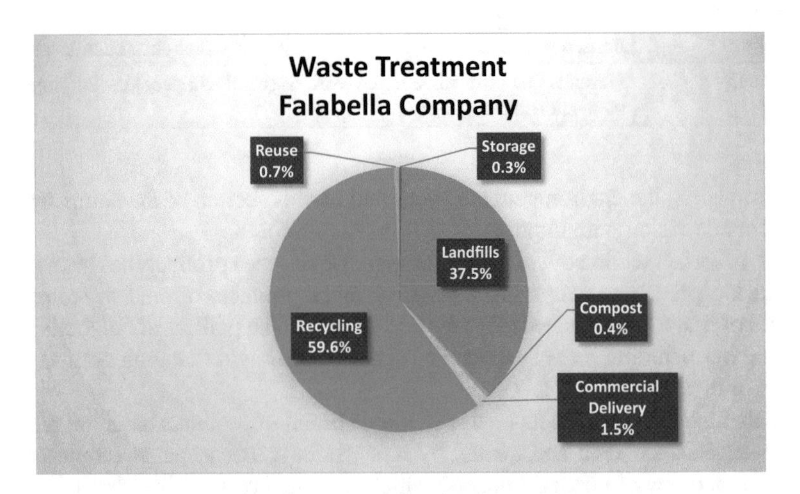

Fig. 8 Waste treatment distribution in Falabella Company [7]

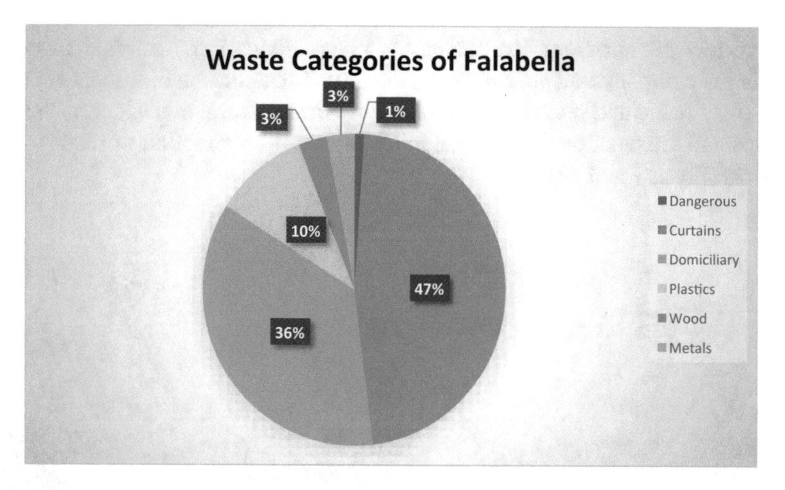

Fig. 9 Waste matters of Falabella company [7]

Table 8 Falabella Retail Companies disposal route, in tons, in 2019 [7]

Elimination type	Dangerous	Non-dangerous	Total
Reuse	0	115	115
Recycling	27	9617	9644
Compost	0	65	65
Incineration	2	0	2
Landfills	0	6083	6083
Storage	55	0	55
Commercial delivery	0	244	244
Total of waste	84	16.124	16.208

4 Retail Companies in Chile: Paris Case Study [8]

Paris is a big company in Latin America, with 44 retail stores with 8518 workers where 70% are women [8]. Paris has a diverse range of suppliers, reaching 1359 in 2019. 58% are national brands and 42% are international brands. 10% are sustainable products out of 30.000 textile products in total.

4.1 Sustainable Program

The sustainability program that is called "Conciencia Celeste" is focused on five areas: better experience for consumers, regeneration of materials, innovative designs, promotion of recycling materials, and donations.

To follow this program, the strategy has been divided in 7 internal sustainable goals in the company that are shown in (Fig. 10). These sustainable goals are related to use sustainable and recycled materials, supporting emergent designers, including organics and artisanal products made in Chile, and the promotion of biodegradable materials (Figs.11 and 12).

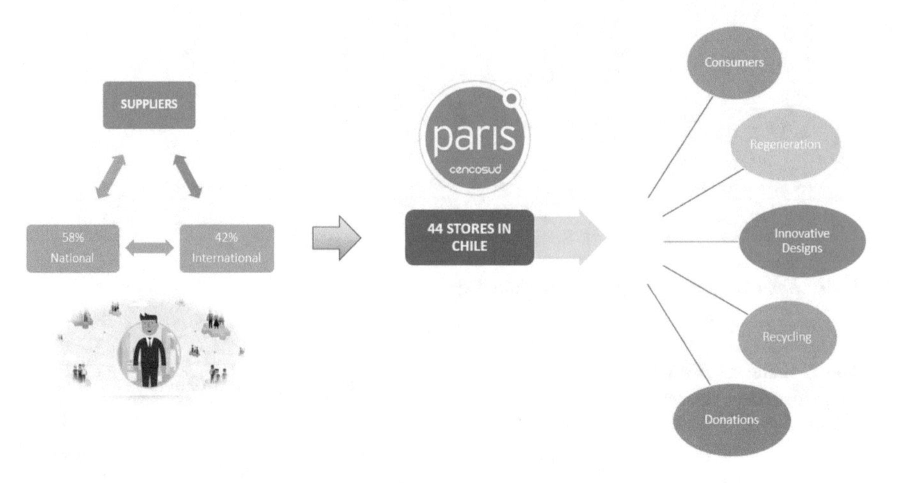

Fig. 10 Sustainable model of Paris Cencosud [8]

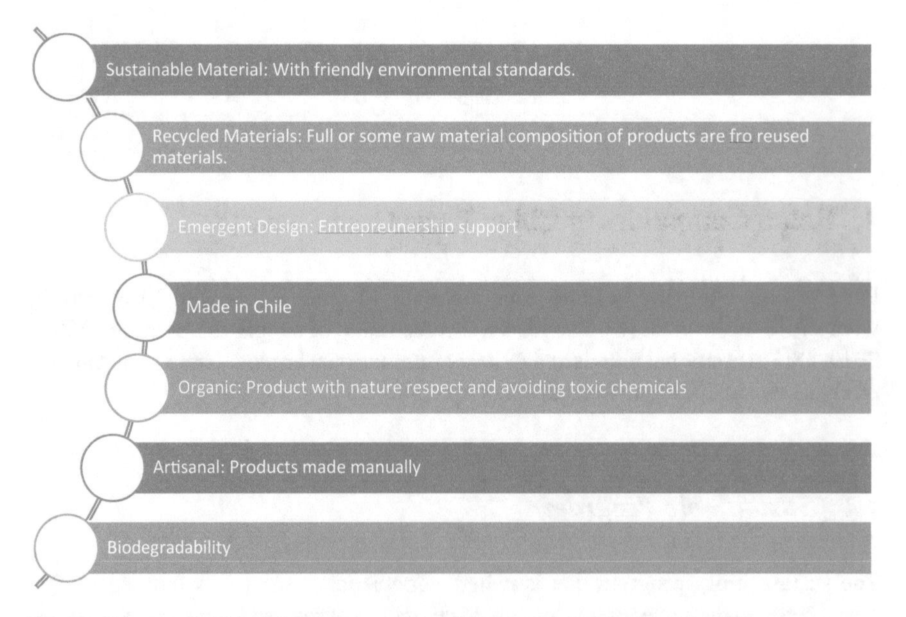

Fig. 11 "Conciencia Celeste" Program and the different sustainable objectives [7]

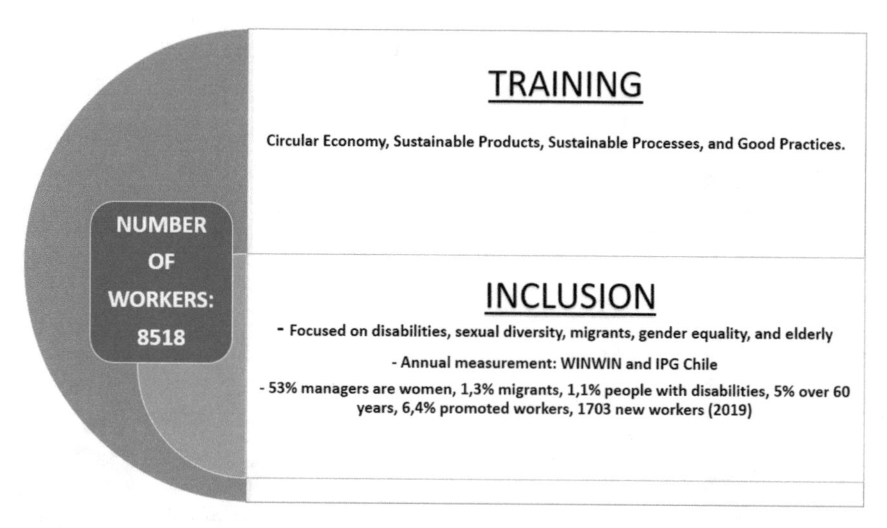

Fig.12 Social Flowsheet Paris Cencosud [8]

4.2 *Social Program*

An important development of programs for workers that Paris company is doing are:

- **Training**: Paris company is working hard to give workers more training of sustainable concepts like circular economy, sustainable products, processes, and good practices with workshops.
- **Inclusion**: focused on people with disabilities, sexual diversity, migrants, gender equality, and elderly people. Paris company does an evaluation process every year that is done with important organizations like WINWIN and IPG Chile.

The inclusion program has worked strongly in including women in high positions in the company, training them with the necessary skills to have successful. Additionally, there has been training for all positions in the company, highlighting the importance of the inclusion and how important is to have a better relationship and environment in Paris company. In a less percentage but not less important, the company is working on the inclusion of migrants in the company, people with disabilities, and people over 60 years old, giving more opportunities to promotion for their workers (Fig. 13).

4.3 *Certifications of Sustainable Products*

Paris has a diverse range of sustainable certificated products, depending on their source materials, principally, like recycled products (plastic bottles, for example), organic cotton products, and wood sources textile product. The objective to certify

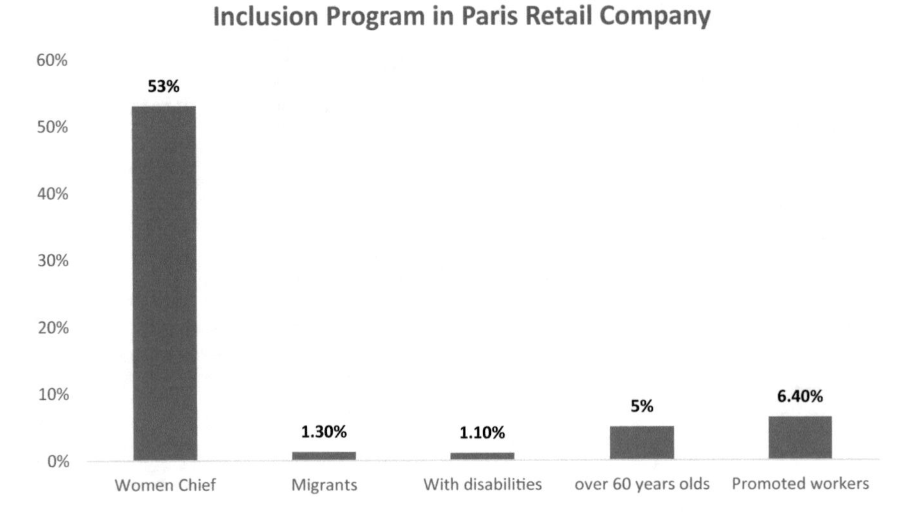

Fig. 13 Inclusion program in Paris Retail Company [8]

their products is a promotion to be more sustainable and be responsible with the environment (Fig. 14).

The principal goal for the company's future is to accelerate the transition to a circular economy scheme, focused on 6 key points: Design, Production, Logistics,

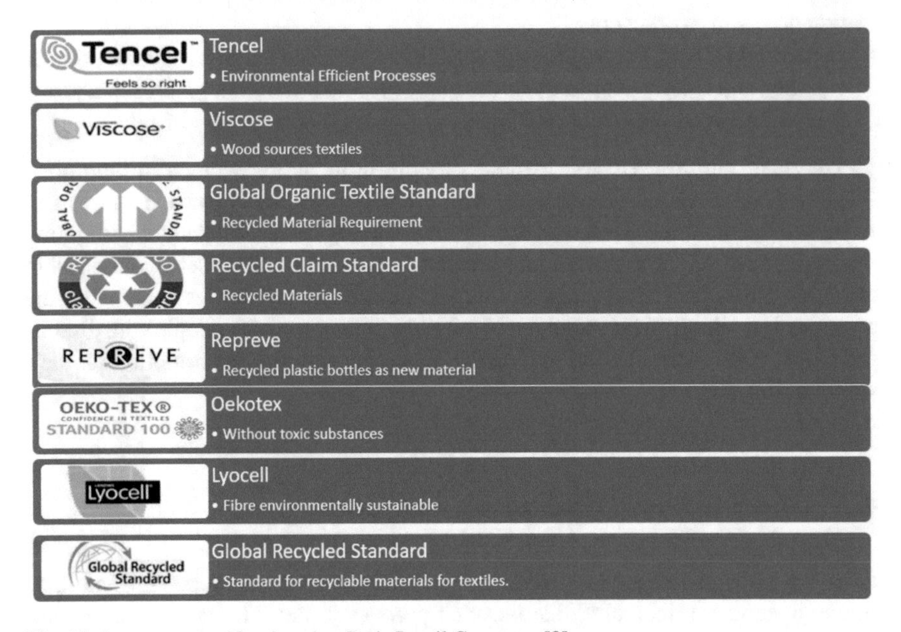

Fig. 14 Important certifications has Paris Retail Company [8]

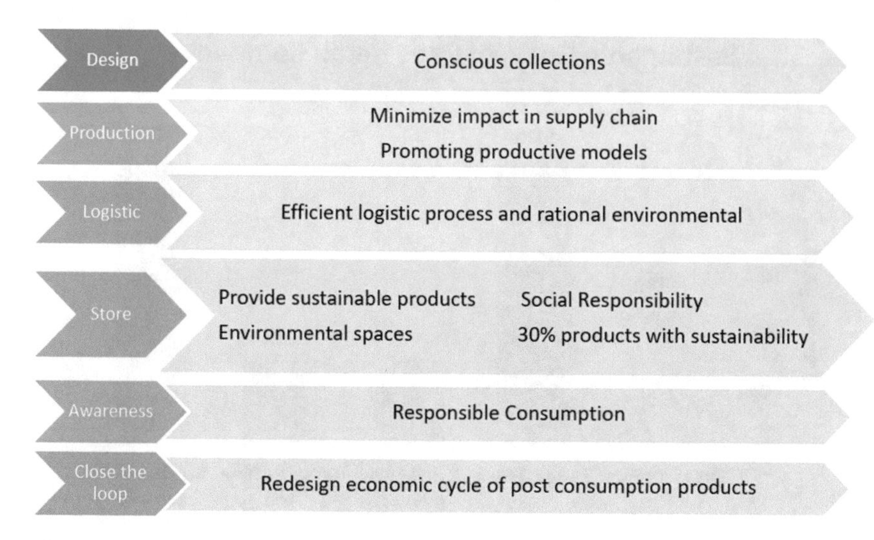

Fig. 15 Traceability of Paris company [8]

Stores, Consciousness, and Close to Loop. Every key point has their own objectives to operationalise the responsible consumption in a cycle loop (Fig. 15).

4.4 Design

Paris wants to create and offer collections that implements the circular economy model. The goal is to reach 100% of sustainable textiles products for 2025. Recently in 2019, Paris has reached 13% of owned brands with sustainable characteristics (Fig. 16).

On this way, Paris company with other organizations and the government have participated in a program called "Volver a Tejer". This project supports artisanal local communities with their designs made of sustainable materials. During their three years of experience, they have supported three versions of the project:

Volver a Tejer I: Support of local communities from the South of Chile. This project was focused on the production of balls of wool, helping 215 women spinners with training in standardization of the product, technical training, quality control management, and machining to increase the quality and productive capacity of the textile product.

Volver a Tejer II: The objective of the second part of the project was to reinforce the relationship between the retail market and artisanal products from Chile. On this time, 13 artisanal groups of workers from the North Chile were included, around 113 spinners, helping them with machinery and training for the workers. Their products are related to the aymara culture, working with the baby category of llama fibres.

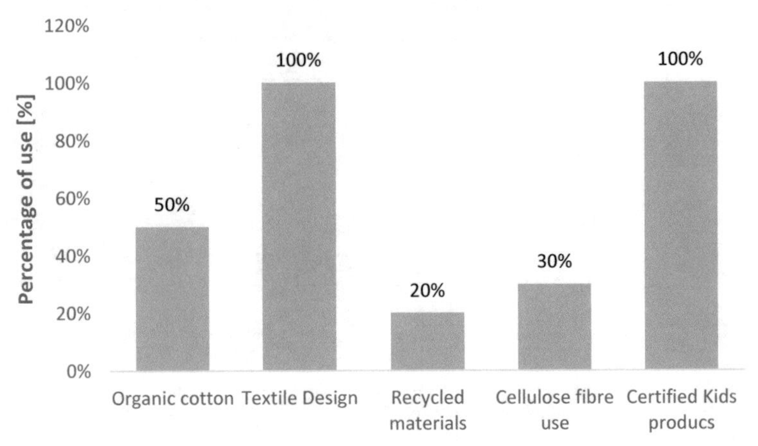

Fig. 16 Details of goals per material in Paris Retail Company to reach a Sustainable Market [8]

Volver a Tejer III: The third part of the project called specially "Vuelve a Tejer con Amor" was the support of communities from Peru and Chile, with the objective to create a fraternity between these groups, sharing their experience with sustainable textile products, in this case alpaca, producing children textile products, a project focused on the social sharing.

The compromise with local communities is a real objective of Paris company. They still want to help the textile development of local communities, supporting in training programs, donations, and a space to sell their products in a big scale as retail company.

4.5 Production

Reducing the impact in the value chain is the goal to implement for a strong plan of traceability of their products, and to have a more sustainable value chain in 2025 [10]. On this way, Paris has implemented a Shared Sustainable program to advance with suppliers, planning strategic objectives to increase the sustainable value chain. The evaluation of sustainable agreements with suppliers have advanced strongly during the previous years, evaluating the traceability and how sustainable are their processes of textiles products that Paris offers to clients in Chile. National and Asian countries especially have been evaluated, reaching 54% of the total of suppliers in 2019. In numbers, this means 76 Asian suppliers and 377 national vendors have developed sustainable agreements with Paris company [10]. The complete plan involves transparent agreements with their suppliers and more sustainable value chains (Fig. 17).

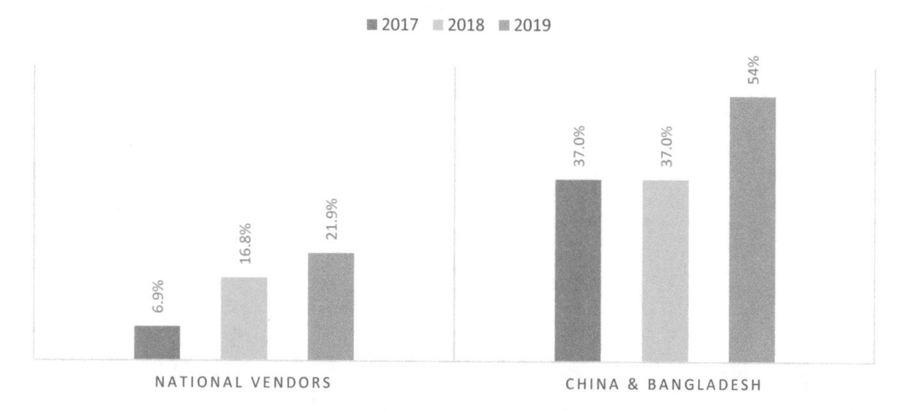

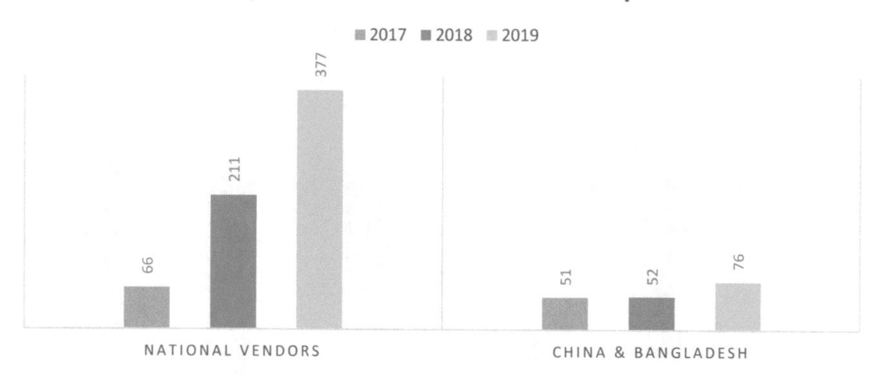

Fig. 17 Sustainable agreements with suppliers from Chile and Asian countries [8]

Other support for small and micro companies' suppliers have been implemented in Paris company. The payment under 30 d and free logistic services for their textile products are some of them, including the promotion of innovative designers in sustainable collections like organic fibres, recycled materials, and tested chemical substances under European norms. Paris has validated 100 styles, 37 approved suppliers, and 50 approval manufacturing (Fig. 18).

The communication with their vendors and the whole value chain is important for Paris company. The creation of the coffee time project is other initiative to have more information about the textile manufacturing processes and a way to manage the compliance of the sustainable program. A software to track their products from vendors is Sedex, a platform that has all the database of the sustainable management of every supplier of Paris company (Fig. 19).

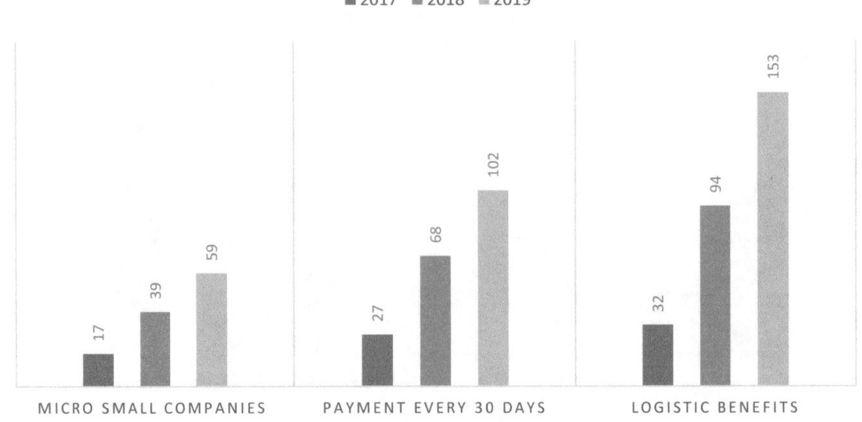

Fig. 18 Analysis of inclusion of micro and small companies, companies with the benefit of payment every 30 d, and logistic benefits by Paris company to their suppliers [8]

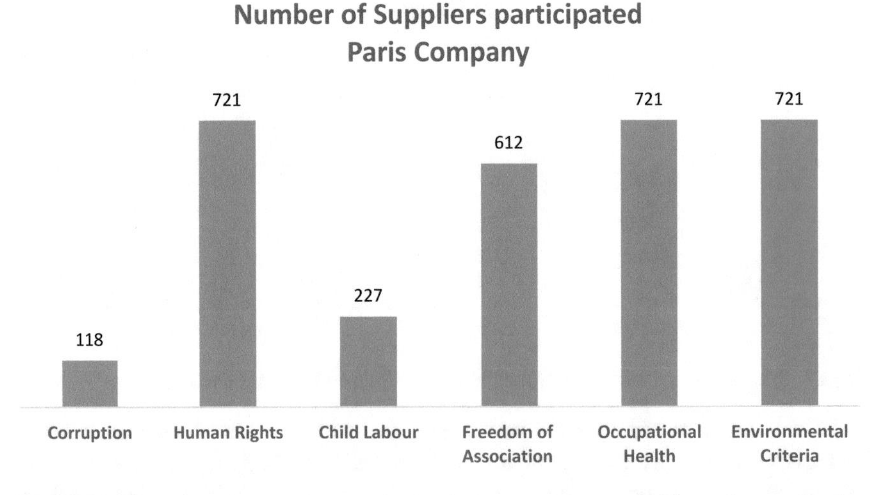

Fig. 19 Number of suppliers participating in the evaluation of Paris, in different themes related to a sustainable development of their businesses [8]

4.6 Logistic

This stage includes an efficient logistic of their processes with a rational environmental treatment. Paris is generating efficient logistic processes with environmental awareness to reach zero waste goes to landfills to 2025. This process involves an

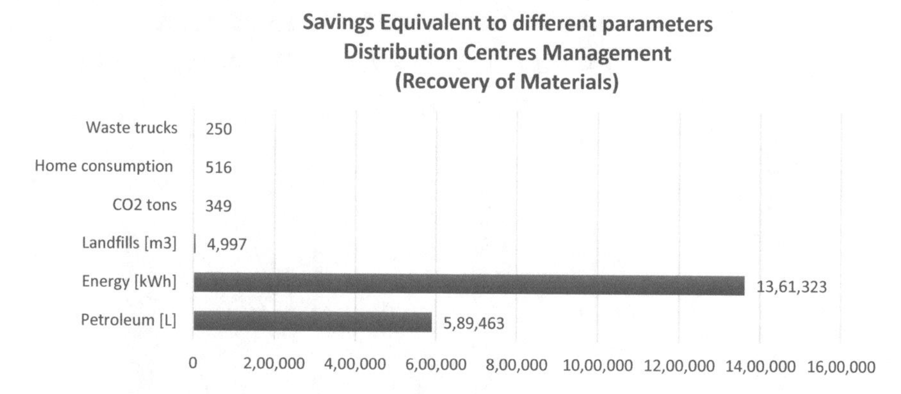

Fig. 20 Savings equivalent to different parameters (recovery of materials) in Paris company [8]

analysis of the principal components of waste generation, creating plans to recycle plastics, cardboards, papers, and other materials.

Other actions are the change of raw material of hangers and alarms of textile products into a more recyclable material (polypropylene) and to standardise and reuse them in every market of Paris retail.

Related to recycling, Paris has a contract with EnFaena, a company responsible for Distribution Centres of textile products. They do the traceability of every material that goes out from Paris markets. The company has saved the equivalent of energy, petroleum, carbon dioxide emissions, and others showed in the next graphic.

Paris retail is changing the logistic of their waste management, where in 2017 100% of their waste volume were destined to the company I:CO in Germany. In 2019, Paris has considered to treat their waste locally in Chile with RETEX company, which last year has covered 90% of the waste volume, reducing importantly the volume sent to I:CO in Germany (Figs.20 and 21).

4.7 *Store*

Paris is increasing the provision of more sustainable products, with special sections related to environmental products, reaching 30% of them with sustainable certifications. Other point is to have more awareness about the social responsibility with their workers, related to their salaries, fair labour conditions as the most important points.

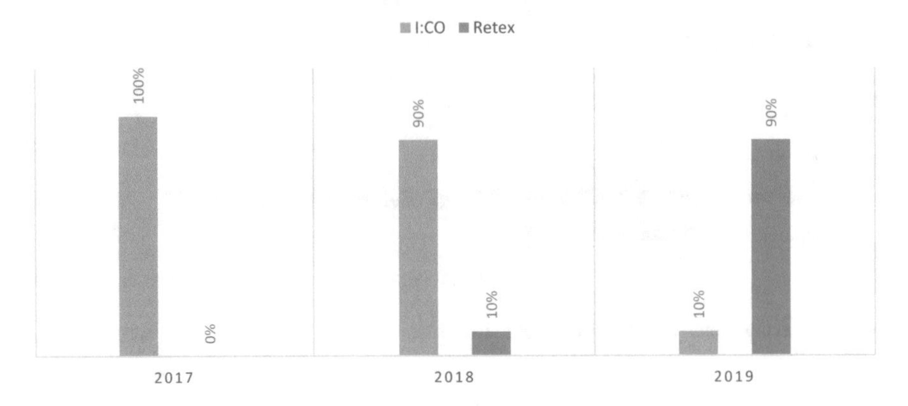

Fig. 21 Waste volume distribution in Paris company [8]

4.8 *Awareness*

A responsible consumption is not only related to the market but with their clients as well. For example, the promotion of sustainable products, services of recycling or reparation of used textile products are some of the projects that Paris is implementing on their stores. For example, a quarter of the total clients of Paris has an attitude for responsible consumption in 2019.

One of the campaigns is to avoid the print of ticket purchase called "I don't print my ticket", with a result of 2 million of paper saved, which represents 17 million of paper metres not printed. 12% of the clients signed this campaign.

"Reducing our residues" is the other program which Paris company is strongly following with the reduction of materials to make a more sustainable vision and behaviour. Paris has transformed the catalogues only via online, avoiding the printing since November 2018. *"Goodbye Plastic Bags"* is the other action which 30 million plastic bags were eliminated.

The last action that Paris is doing on its markets is the elimination of 70% of publicity on streets, helping in the reduction of PVC residues, having an impact of 91% recycling.

4.9 *Closing the Loop*

The main objective of the sustainable program is to close the loop, following the destination of their products (post consumption) and how they can create new actions to redesign waste textiles, transforming the system economically. The goal is to triplicate the reinsertion of products and it is managed with 4 important actions:

Waste Management: The company responsible is Modulab, which reuses Paris publicity recycled materials to make new products like wallets, reused bags, etc. focused on the eco-design. With this action, the recycling of publicity from streets reached 91%, where the company has recovered 5544 m^2 of PVC, which means 1770 tons.

"Ropa x Ropa Program": This campaign has a long story since 2013. This is related to the recycling of worn clothes in Paris market, an activity for clients for free, giving them a 30% discount to buy new products in the shopping company. Tons of textiles recycled have been 982 tons in 7 years.

"Reparalab Program" was born with the aim to recover and having a second use to old textile products of clients, offering a free clothing repair service to extend the product's life. This program is supported by the Environmental Department of the Government.

"Love Demonstrations" is the last actions offered by Paris, where new clothes are sent to organizations to give these products to people with needs.

In conclusion, Paris has a hard goal to reach 100% of sustainable products from their own brands in their stores in 5 years from now, where the company is advancing fast including new sustainable products, demonstrated with the support of Volver a Tejer projects for example, and the traceability plan with their suppliers to include sustainable products on their market. The company has a big challenge until 2025 to comply with every objective that they have written related to a sustainable model in the company.

5 Retail Companies in Chile: Ripley Company [9]

Ripley has 46 Stores in Chile and controls other big retail stores in the country. The company has its own sustainable line that promotes an environmentally friendly change in the company with the use of sustainable raw materials.

5.1 Sustainable Line

Ripley has recently created the Respect Line, a space that offers sustainable products with different source of raw materials such as natural fibres, organic cotton materials and products with less water consumption are their key to reach a sustainable market (Table 9).

Table 9 Sustainable products promoted in Ripley stores and their principal certifications [9]

Product Type	Description
Natural fibres	Without the use of synthetic chemicals
Organic cotton	Cotton without fertilizers
BCI certification	Cultivation and harvesting with respect of the planet and with high quality of workers involved
Water Uue reduction	Products with 50% less water consumption
OEKOTEX certification	To guarantee products free of toxic chemicals

5.2 *Sustainability with Ripley Workers*

The development of sustainability in their workers in Ripley are related with leadership training, integration of workers with disabilities and alliances with foundations that supports the diversity. *"Iguales"* foundation and Inclusive Network with Sofofa are some examples. Additionally, Ripley has awareness for the satisfaction of workers with the company, creating a perception test about labour conditions in Ripley. Last results from 2019 have demonstrated 90% of workers never felt discriminated in the company.

In energy consumption, Ripley has signed an agreement with Colbun company to work to reach 100% renewable energy supply, which Ripley has started to use renewable energy in 27 stores. The reduction in using renewable energy in the company is the equivalent of reduction of 11.250 vehicles per year.

5.3 *Recycling Projects*

About recycling, Ripley has worked in two projects: **Donation and Recycling.**

Ripley donates every month clothes in a second-hand to Coaniquem, Debra and Hilo, social organizations to help people with inadequate resources. In 2019, for example, more than 27,000 units were donated.

In recycling, Ripley has implemented a pilot project to recycle three kind of materials: plastics, paper, and aluminium cans. However, the most important project is the fibres recycling to transform materials into a new product or yarns, having an alliance with Ecocitex, a recycling company that follows a sustainable model on its business, managing the second hand and waste textile products of Ripley and other textile sources.

Other sustainable objective for the company is the promotion and support of entrepreneurship, offering them the chance to show their products, supporting the national market. This action is joined with Commerce Camera of Santiago.

In conclusion, Ripley has its own line of sustainability in the company which most of them are related to the transformation of new materials, donation, support of entrepreneurship and its owned workers with leadership training. However, based in the literatures found, there is a lack of information about volumes, and goals to reach the goals of sustainability in the country. It would be interesting to know what the volume is recycled to create a textile value chain of the products.

6 Recycling Textile Companies in Chile

It is extremely difficult to trace that amount of textile manufactured, imported, sold, discarded and recycled in the country since there is not enough information available to make an efficient and precise value chain of textile products. However, there are projects that aim at recycling different textile products, creating agreements with big sources of textiles such as retail stores. On this chapter, three examples: Ecocitex, Travieso, and Retex are reviewed.

6.1 Ecocitex

Ecocitex commits to raising special awareness in following a circular economy model. The company works under sustainable principles, supporting ethical **businesses** fair working and salary conditions, targeting at economic, social, and environmental impact. All recyclables are proccessed and not recyclables are transformed into **yarn skeins** [10]. Part of the textiles come from Ripley and Falabella retail companies. Then, Ecocitex processes the matters and redesigned products are sold in Falabella and Ripley markets as sustainable products. Other source of textile is Travieso, a company that receives direct donations from clients, which are sent to Ecocitex afterwards [11].

In 2020, this company has increased its capacity to recycle up to 4200 kg of clothes per year. Their goal the future is to produce 5 tons of yarns per year; hence, the company needs a strong strategy to collect more second-hand textiles as raw material to reach this ambitious milestone [10].

Also, Ecocitex has a social development mission that intends to include women deprived of freedom and violence victims in its workforce, by partnering with Abriendo Puertas Corp in order to integrate these workers into labour and under fair conditions [11], as a means of recovering from their traumatic experiences. Other social initiative is the exchange of clothes with Maria Ayuda Foundation, offering clothing in good conditions to people in need.

6.2 Ecofibra [12]

Ecofibra is a company that works recycling of second-hand textile products and discarded textiles so to give them a second life use, receiving 15 tonnes per day of second-hand textile products in the north of Chile, specifically in Iquique. Ecofibra receives textile products from ZOFRI store, one of the biggest dutty free companies in Chile that imports textile products from Asian countries. Ecofibra has been able to recycle 39,000 ton of residual textile, representing 66% from the total of textile products. This recycling process avoid the incineration of textile products and landfills destinations. Ecofibra transforms recycled textile materials into isolated board panels, which have a price 65% less than mineral wool material, giving a second opportunity to disposed textile material, improving its value chain and cycle of use.

6.3 Retex [13]

Retex is a company that is responsible of recycling textile products from Paris retail company, avoiding the delivery of clothes to I:CO company in Germany, that was in charge of the recycling process of Paris company previously.

In Retex they have seen in COVID-19 is an opportunity to create new accessories from their recycled sources, creating an effective system, by transforming products into new ones. As an overview, the process consists in receiving the recycled textile materials, sorting in two categories: denim, cotton, and other textile sources (Fig. 22).

- Denim is transformed into coat racks and other personal articles.
- Cotton and other textile sources are transformed into reversible face masks.

Other Retex projects, using second-hand textile products as raw material, are:

- Reuse of garment in good conditions.
- Production of cleaning wipes.
- Manufacturing of acoustic panels.

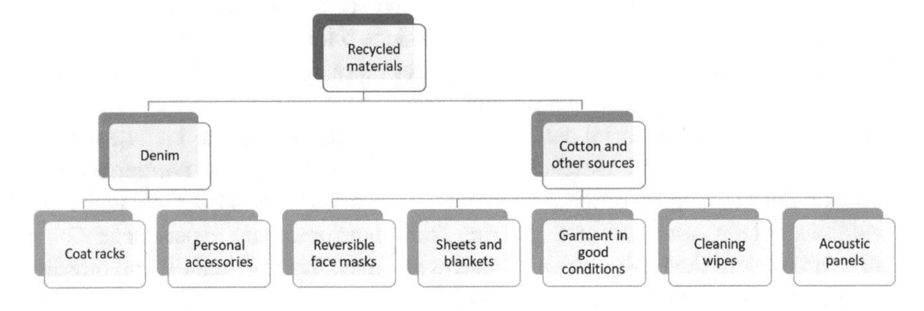

Fig. 22 Production scheme of Retex company [13]

6.4 Travieso [14]

Travieso is a company founded in 2018, focused on recycling specially kids' clothes and transforming them into new textile products. Its headquarters are located in Santiago with salerooms available. The vision of the company is aligned with three sustainable goals: Reduced inequalities (#10), Responsible consumption and production (#12), and Climate Action (#13).

Travieso company receives kids' clothes from different sources and has three options for clients: The sale of second-hand kids' textile products and accessories, reception of second-hand kids' textile products with the option of credit or discounts for Travieso products, funds transfers, or kid's clothes donations to the company for the transformation of new products or new raw material. Travieso has multiple salerooms in Santiago for kids' clothes. However, due to COVID-19 crisis, this system is in standby until resuming activities is possible.

7 Analysis of Circular Economy in Chilean Textile Market

The work discussed above has shown a first overview of the Chilean textile market, with special attention to the three most important retail companies in the country. The work was based on information from sustainability reports published in 2019 which have demonstrated that every company has set the goal of becoming more sustainable in the future and to incorporate circular economy principles in their businesses.

These three companies have a solid program that include their workers and clients in order to offer the best experience buying in their stores. The use of technology has been important during past years, where online platform is introducing new clients and getting them closer to the store and their products in a different way. The automation of some processes has permitted the fast and personalised attention of clients, reducing dead times in stores. In this way, other project like the elimination of printed tickets and plastic bags have permitted to save tons of paper used unnecessarily and eliminating the use of plastic bags as a packaging product. The importance of the last project is in alliance with FSC to eliminate as many plastic bags as possible in retail companies, forcing them to find new ways of packaging, like paper bags that have been increasingly used in retail companies lately.

Regarding to their workers, every company has proposed to implement almost the same strategies, such as inclusion, integration of elderly people, women, migrants and young professionals to the workforce of companies in order to give similar opportunities for men and women, aligned with one of the sustainability goals 2030. Awareness of workers satisfaction related to labour conditions has been an important topic since the implementation of surveys for workers in order to know their opinion about the conditions in different working positions. This action has allowed to raise awareness about working conditions and to highlight important things for workers so that companies can offer a better work place. The increase of salary for workers has been other important action that have satisfied workers having a stronger commitment to their companies. The minimum salary in the companies is 1.8 times the minimum salary by law in Chile, translating into more benefits for workers and their families.

Retail companies are advancing in incorporating exclusive areas to display sustainable products. Every retail company analysed in this work has created a new brand to promote sustainable products, with the purpose of keeping it apart from the regular textile products. Retail companies mean to move forward by developing a bigger sustainable market to reach a 100% sustainable materials in their own brands in the future. By successfully implementing this strategy, the pressure on landfills would dramatically decrease due to the high biodegradability quality of sustainable products and higher recycling rates. Additionally, the suppliers from Asian countries would also have to improve the biodegradability rate of their textile products in the market, demanding them to be more responsible with the environment and transforming the textile value chain into a more sustainable model.

The presence of local communities in textile market is increasing annually. Retail companies have paid attention to the demand of artisanal products in the country, offering in exchange training programs like quality control, and new technologies; and the opportunity to sell their products in sustainable lines on retail market. It is estimated the presence of local communities will continue to grow in the future, incorporating new artisanal designers who works with sustainable materials like wool, alpaca, and llama, national products with a high quality and more environment friendly processes.

Retail companies have an alliance with important recycled companies in Chile during the past years, with the aim to increase the recycling rate of their textile value chain. Most of recycled companies are created by entrepreneurs who have found an opportunity in the high amount of discarded textiles in the country, transforming them into new sustainable materials and products, thus, expanding the business of textile industry in Chile. For instance, Ecofibra isolation panels made by second-hand textile products, Ecocitex yarn skein, Retex personal accessories and a second-hand products have been the best seller ideas by these recycling companies, supporting the sustainable market, and adding value to a disposed material that, otherwise, represents a contamination source.

The association with retail companies that closes the loop of textile cycle is advancing slowly and self-governed in the country. However, association with textile trading companies such as retail is increasing the recycling rate in Chile. Nonetheless, these companies need to grow faster in order to keep up with their agenda, since their goals are due by the fast approaching 2030. The support provided by the government and private companies would be key in accomplishing the big change needed to close the gap and to have a much desired textile Circular Economy model in Chile (Fig. 23).

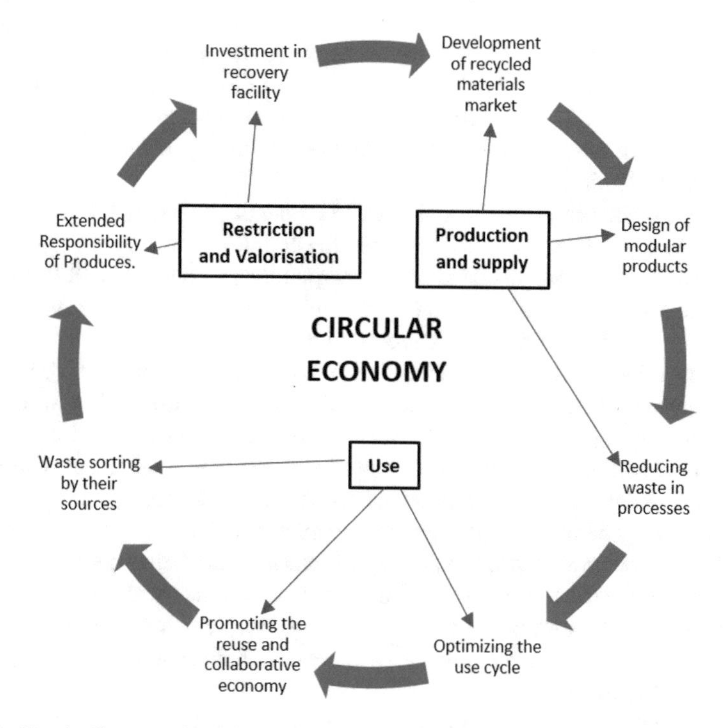

Fig. 23 Circular Economy Model in Chile [3]

8 The Circular Economy Route of Chile's Proposal

Out of the total of imported textiles in Chile, it can be estimated that 73% goes to landfills [1], and the other percentage is distributed into recycling (20.4% estimated) and other unidentified ends. If numbers are compared with the volume of sales, the recycling percentage is too low as to have a big impact in the textile value chain. Instead of that, retail companies are improving their conditions and increasing the volume of recycling annually, in association with other specialized companies dedicated to recycling, with three different approaches:

- Second-hand market from used textile recovery, creating a new low-price market.
- Transformation into new materials from unworn and worn clothes (i.e. skein from recycled textile products).
- Transformation into new products with a high range of varieties.

Retail companies are strongly concerned about traceability, a topic that was criticized in the analysis of REP law in Chile. The evaluation of Asian textile products in retail companies is a big step forward to reaching a sustainable market, since it is this kind of products that represent the biggest volume of sales in the country at every scale of market.

The current landfills in Chile have an approximate lifespan of only 12 years. According to this estimation, Chile is collapsing under its waste. It is an urgent problem that must be managed accordingly in order to avoid an environmental catastrophe, and it can only be done so by aiming tat drastically reducing the amount of waste reaching landfills.

The solid waste production in Chile is 535 kg per capita yearly [3]. This shows it increased during the previous years, leaving the country worryingly over the OCDE average. In 2000, Chile generated 295 kg/capita yearly, but by 2018 the country reached the average of 400 kg [3]. Sadly, the solid waste recycling has reached only 2% in Chile, which is far from OCDE countries [3].

Citizens' minds have changed about their relationship with shopping. They question companies about the materials used and push them into staying accountable for their marketing models in topics such as the use of chemicals, local production, reuse, recycling programs for their products and energy efficiency. Customers are currently more informed, empowered and have a higher environmental awareness than older generations. This trend was demonstrated during the social outbreak in 2019, when people from Chile demanded to live under better conditions. Retail marketing models were strongly questioned then due to many reason that included their role in the deterioration of the environment. People demanded the markets to change into more sustainable, more transparent in their data and to have a better traceability of their products.

The Circular Economy transformation is expected to create new job opportunities in the country, a side benefit that promises to improve employment rates in the country with openings in areas such as waste management, restoration and product maintenance, and specialized professional services.

Chilean government has implemented two periods of time to comply with the Circular Economy agenda: One period is between 2021 and 2030, and the other until 2040 (Table 10).

Considering that the textile industry is one of the most contaminating in the country, this plan needs to be implemented with all the actors involved in the textile

Table 10 Circular economy agenda goals in Chile [3]

Parameter	2030	2040
Solid waste generation	10% of reduction	25% of reduction
New employments	100.000 new employments	180.000 new employments
Solid waste recycling	Has reached 30%	Has reached 65%
Waste generation per PIB	Has reduced 15%	Has reduced 30%
Material productivity	Has increased 30%	Has increased 60%
Recycling rate	Has increased 40%	Has increased 75%

industry because these goals also aim at reaching a sustainable economy. It seems that the value chain of textile products is improving mostly thanks to the private sector, with only partial support of the government in particular actions. It is important that private sector and the government align goals, as to work together for better industrial conditions.

9 Conclusion

When it comes to creating new ways of becoming a more sustainable business, transparency regarding stock taking is key especially in textile market. Since it is this information the one that would allow us to make Circular Economy maps. With the information provided in this research, an initial mapping of textile products was designed. Unfortunately, a complete map of textile products in Chile is not possible in a 100%, due to lack of access to information about volumes managed by retail companies nationwide. However, with the information gathered here, there was an estimated Circular Economy map created to show the value chain of textile products in Chile including the most important retail companies, partnerships with recycling companies and other projects that allow us to a better view of the recycling prospects in the country (Fig. 24).

The decreasing of this 73% of textile products that reach landfills is the biggest goal for the textile market in the country. Recycling companies have committed to reduce the volume of discarded textiles in Chile, giving them other destinations, and increasing the recycling rate of the market.

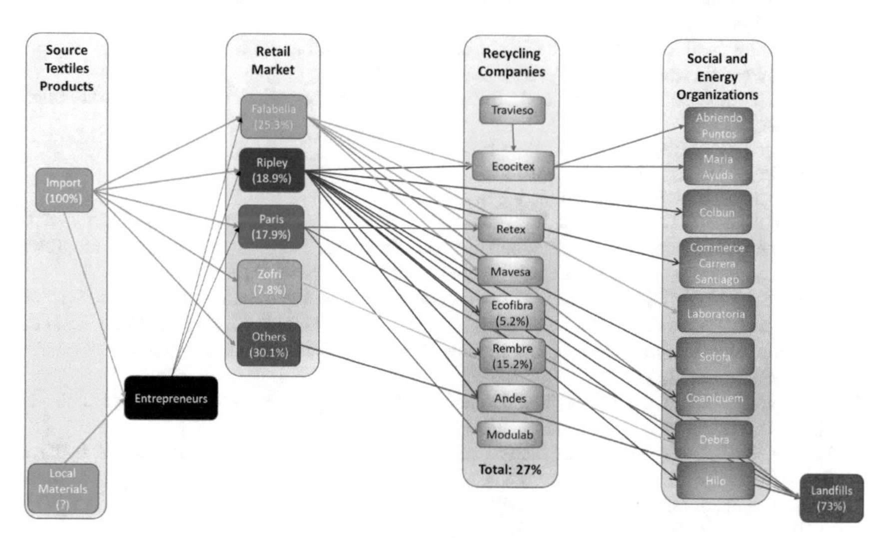

Fig. 24 Map of textile industry and the estimation of textile volumes in Chile

Recycling companies are partnering with retail companies to help each other with the managing of disposed textile materials. Currently, the most influential companies are Ecofibra, Ecocitex, Retex and Rembre due to their connections with the most important retail companies. These partnerships allow recycling companies to receive higher volumes of discarded textile and to start a bigger recycling rate and the retail companies to widen their sustainable brands. These alliances should be supported by the government, who, by doing so could require tracing information of the stock taking, which is key to determine the real impact retail stores have in the value chain.

In conclusion, the current recycling rate in the country is still too low compared with other OCDE countries. This translates into a small impact in the decreasing of textile volumes reaching landfills. However, new strategies that come from within retail and recycling companies are expected to improve the scenario by increasing recycling rates, improve traceability of imported textile products, and including more local products into the retail market. This will allow further construction of the value chain of textile industry in Chile which will help to determine more efficient solutions for the development of a sustainable system with a Circular Economy model.

References

1. Monroy M (2020) Viste Consciente. Chile University, Architecture and Urbanism Faculty
2. La Tercera (2021) Ecocitex: Eliminar el desecho textil en Chile. https://www.latercera.com/masdeco/ecocitex-eliminar-el-desecho-textil-de-chile/. Accessed 6 Feb 2021
3. Ministerio del Medio Ambiente, Chile (2020) Hoja de Ruta Nacional a la Economía Circular para un Chile sin Basura
4. Garcia I, Muntal A (2017) El mercado de la confección textil y el calzado en Chile. España, exportación e inversiones
5. World Integrated Trade Solution (2020) Textiles and clothing export and import by region. http://www.wits.worldbank.org/. Accessed 1 Feb 2021
6. Vidal C (2020) Fashion, law and moda. https://enriqueortegaburgos.com/industria-moda-chile/. Accessed 30 Jan 2021
7. Falabella (2019) Reporte de Sustentabilidad
8. Paris (2019) Paris Conciencia Celeste
9. Ripley (2019) Resumen Ejecutivo, Memoria Ripley
10. Pascual I (2020) La hazaña de Ecocitex, la empresa B que busca acabar con los residuos textiles en Chile. https://www.janegoodall.cl/post/la-haza%C3%B1a-de-ecocitex-la-empresa-b-que-busca-acabar-con-los-residuos-textiles-en-chile. Accessed 30 Jan 2021
11. Diario Sustentable 2020 Ecocitex: Economia circular textil que transforma ropa en desuso en hilado reciclado similar a la lana. https://www.diariosustentable.com/2020/04/ecocitex-economia-circular-textil-que-transforma-ropa-en-desuso-en-hilado-reciclado-similar-a-la-lana/. Accessed 30 Dec 2021
12. BiobioChile (2020) http://Reciclaje: de residuos textiles a paneles aislantes. Accessed 30 Jan 2021
13. Pulso (2020) Empresa chilena fabrica escudos faciales, mamparas y mascarillas a partir de residuos reciclados. https://www.latercera.com/pulso/noticia/empresa-chilena-fabrica-escudos-faciales-mamparas-y-mascarillas-a-partir-de-residuos-reciclados/BNIVGA5YKFD3HERCXTH7T6FAB4/. Accessed 6 Feb 2021
14. Gonzalez C (2019) La empresa Travieso se inspira en el trueque para reutilizar el 100% de la ropa de guagua. https://www.paiscircular.cl/consumo-y-produccion/la-empresa-travieso-se-inspira-en-el-trueque-para-reutilizar-el-100-de-la-ropa-de-guagua/. Accessed 14 Feb 2021

Circular Economy in Textiles and Fashion

Shanthi Radhakrishnan

Abstract The production and manufacture of the textile and fashion industries are known for high utilization of natural resources and the impact on environment and community. The economy that was linear (take-make-waste system) has shifted to a circular economy where product designs are created to reduce waste and the waste generated is converted into new products (upcycling) or resources (down cycling) which go into the second generation of the raw material cycle either in full or as a percentage of the whole. Over the years, the global textile industry has grown to higher levels in terms of production and consumption, but has created a whole set of issues due to the linearity in production and economy. Now efforts are being directed to reduce the waste and close the loop for zero wastage creating a whole lot of challenges and opportunities that is translating itself not only in the manufacture and production techniques and standards but also in the marketing and advertising campaigns. Leading global brands like NIKE, Patagonia, Speedo, BASF are announcing their pledges and statements toward their contribution to circular economy and targets to be achieved in the near future. The new journey of the textile industry is to explore the methods of incorporating the principles of the circular economy in all levels of product development and stake holder consciousness. Some critical points of consideration are the creation of new business models that increase the utility of clothing, inclusion of safe and renewable inputs and smart solutions that convert used clothes into new ones. Similarly many solutions may be available or may be developed, but manufacturing and industry must aim at zero waste production and circular economy.

Keywords Circular economy · Responsible production and consumption · High value recycling · Closing the loop · Zero waste

S. Radhakrishnan (✉)
Department of Fashion Technology, Kumaraguru College of Technology, Coimbatore, India

S. S. Muthu (ed.), *Circular Economy*, Environmental Footprints and Eco-design of Products and Processes, https://doi.org/10.1007/978-981-16-3698-1_6

1 Introduction

The production and consumption of goods and services coupled with finance contribute to the economy of a state, region or country. Whatever be the economy whether Feudalism, Socialism, Communism or Capitalism, the type of economy is determined by the respective nations on the basis of population, the resources and wealth distribution and the cultural background of the nation with the focus on social upliftment and progress of the nation. Nations have been driven by money, wealth, production and consumption patterns (economic focus), but the challenge is the transformation of the global economic systems to sustainable ones. If we look at nature, it is evident that there is a limit that governs the goals of every species to ensure compatibility with the restraints of the ecosystem in which they thrive. This phenomenon guarantees the feasibility and sustainability of the system [1]. If this principle is reflected, while planning the framework of a new economy, long term benefits and economic gains will result.

The governments all over the globe are looking out for an economy where resource consumption and waste accumulation are curtailed by different means like sustainable design/remanufacturing/reuse/recycling (circular economy) in contrast to the 'take-make-dispose' model of production (linear economy). The second law of thermodynamics states that when energy changes from one form to another or matter moves freely, entropy in a closed system increases [2]. It has been reported that the quality of materials and energy reduces when extracted, used in a limitless manner and converted to various forms, due to the increase in entropy. Entropy is a measure of molecular disorder or randomness of a system. Let us take an example: if one kilo of gold is made into an ornament, it is more valuable when compared to the gold which is distributed over microchips in mobile phones. Recovering the gold from mobile phones is a humongous task which causes quality and money loss and this loss has to be compensated with a new input of raw material. Circulation and reuse of matter and energy through the economy helps to decrease the new input of raw material and delays the speed of entropy [3].

In 1989, two environmental economists David W. Pearce and R. Kerry Turner pointed to the world that the traditional linear economy would not help and that the environment had become a waste bin. Their study was based on living systems which were nonlinear, but taught many lessons about effective survival techniques. It focused on the optimization of systems rather than components and on design for fit and function. They highlighted that the **C**ircular **E**conomy is based on principles like removal/reduction of waste and pollution, maximum utilization of products and materials and thirdly regeneration of systems in nature [4].

Circular Economy addresses two phases namely the biological phase and the technological phase. In the biological phase, biology based materials like cotton or wood can be made to enter the parental system by composting or anaerobic digestion. This results in regeneration of the soil, which is the existing natural system, thereby enhancing the renewable resources [5, 6]. In the case of technological phase, the application of scientific knowledge leads to recovery and refurbishment of components/materials that is made possible by the adoption of strategies like the 6R's in sustainability [Recycle, Rethink, Refuse, Reuse, Repair & Reduce] [7, 8].

1.1 Concepts Leading to Circular Economy

- The functional service economy or performance economy
  Economy has been categorized as the Stone Age economy, Industrial Economy and the Performance economy. Goods or products have been classified as bulk goods and smart goods depending on the value it fetches in par with minimal use of resources. Bulk goods transactions were found in the Stone Age Economy and in the Industrial Economy; smart goods transactions have been carried out in the Industrial Economy and the Performance Economy. Bulk goods are those which are stored, transported and sold in large quantities without any packaging, but are considered low in value, e.g., sand, gravel and coal; smart goods are high-value goods sold in small or niche quantities, e.g., medicines and enzymes.
  In the Industrial Economy, the sale of goods have the risks or liability for quality and utilizations to be borne by the consumer or the service provider. This phenomenon is known as 'externalization of risks'. In the Performance Economy, the cost of risk or liability has been shifted from the consumer to the producer known as 'internalization of risks'. The added feature is economic incentives for loss prevention and waste prevention. The Performance Economy also works toward job creations and better utilization of resources. This can be achieved by extending the service life of products, conservation of materials, substitution of renewable resources with non-renewable ones. Thus, the main aim of the performance economy is to increase wealth and job creation with decrease in resource consumption when compared to the Stone Age and Industrial Economy as seen in Fig. 1.
- **Cradle to Cradle design philosophy**: The resources in nature are created with the philosophy of disassembly and reuse to return back to the soil or parent population

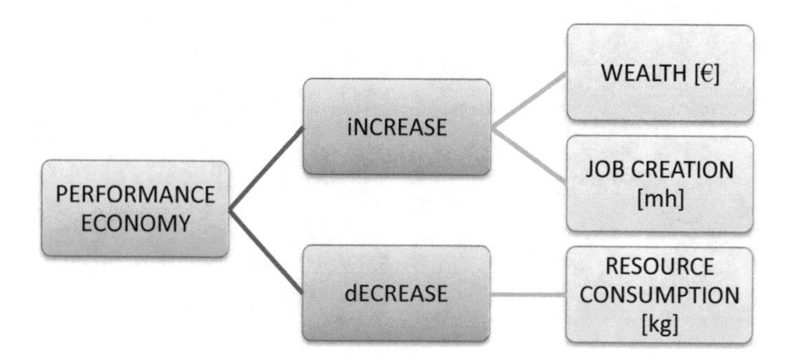

Fig. 1 Objectives of Performance Economy [9]

either as biological or technical nutrients or resources. Any organic material that can decompose into the natural environment and serve as food for microbiological life thereby preventing pollution is considered as a 'biological nutrient'; inorganic or synthetic materials that have been manufactured, e.g:., plastics, can be reused many times without/minimum loss in quality as a continuous cycle are termed as 'technical nutrients' as seen in Fig. 2. The lessons learnt are that any matter in nature becomes a resource for some other matter. Further, living organisms are dependent on solar energy to thrive showing that the human race should use clean and renewable energy like geothermal, gravitational, solar, wind and other forms for any development; processes in nature. Like geology, hydrology, nutrient cycling and photosynthesis respond to each environment across the Earth showing that the designs created fit appropriately to each locality and revealing the efficiency and perfection in design creation. These concepts in the Cradle-to-Cradle principle highlight the opportunities for design innovation and creativity with positive impact on society and environment.

- **Biomimicry**: Many living organisms found in nature face the same challenges as human beings, but they are able to meet them by sustainable means namely handle forces, move water, create color, safe energy distribution and fastening things together. Nature can serve as a model, as a measure and as a mentor as it has a vast chest of knowledge embedded in it. Biomimicry is a tool in design creation that emulates the strategies used by nature or living beings. Development of forms, processes and eco systems can be done with the help of design inspiration from nature. The humpback whale's flipper has bumps on the leading edge which has become an inspiration for designing blades the in wind turbines and airplanes; the whale fin inspired blade design fetched the same amount of power at 10 miles per hour as the conventional blades captured at 17 miles per hour [11,

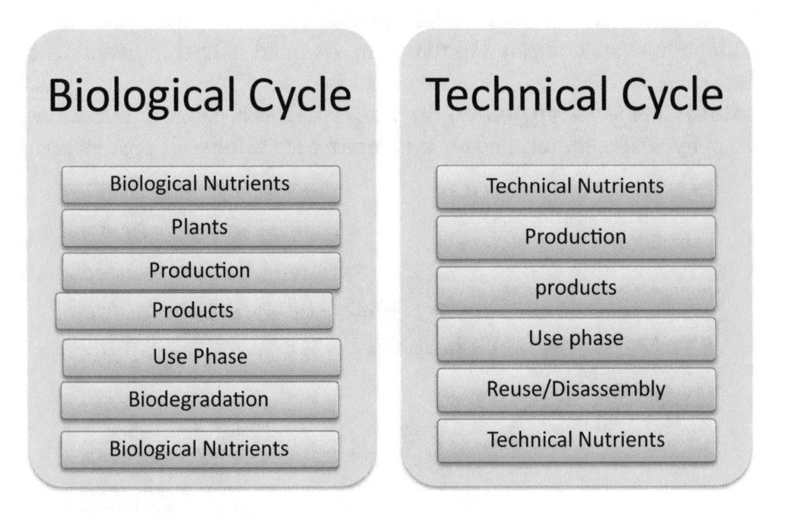

Fig. 2 Nutrient metabolism [10]

12]. Moreover the swirling vortices created by these blades provide extra air lift thereby reducing noise and increasing efficiency by 40% [13]. Similarly there are processes or whole eco systems which may be biomimiced. Other examples include train front design from the Kingfisher bird, lotus leaf cleansing for the new generation paints, anthill structure [termite] for sustainable buildings, Tsunami warning devices from dolphins [14, 15]. Kevlar is a strong manmade fiber that has high strength and modulus and thermal stability. Spiders in the forest produce 'Spider Silk' that is much tougher than Kevlar and equal in strength and does not need any resources required for the manufacture of Kevlar, but uses 'digested flies'. Similarly, plants produce cellulose that is stiffer and stronger than Nylon and also creates a matrix in wood—a natural composite that has higher bending strength than steel [16].

- **Industrial Ecology** This approach deals with three segments namely the flows of materials and energies for industrial and consumer activities; the impact of these flows on environment and the economic, political, social and regulatory influences on the flow, use and conversion of resources [17–19]. The external inputs [materials and energy] and the release of wastes to the external environment are important features and based on this the eco systems are classifies as Type I, II & III. The Type I eco system is linear and highly dependent on external resources with unlimited waste; Type II eco system is partially cyclic using limited energy and resources with limited waste; Type III eco system is cyclic and highly dependent on recycling and reuse resulting in zero waste as seen in Fig. 3. The core elements of this system include system perspective in environmental analysis and decision making, innovation and technological change to enhance eco design development, role of business organizations to implement and experience the impact of industrial ecology, moving toward dematerialization and eco efficiency, analysis and evaluation of the changes incorporated in the eco design and processes and integrating all the elements into the eco system at all levels for sustainability.
- **Natural Capitalism**: At the start of the Industrial Revolution, there was an abundant availability of natural resources but labor was scarce. In current times, the situation has become reversed with surplus people, but there is scarcity of natural

Type I Ecology	Type II Ecology	Type III Ecology
• Linear Material Flows • Use of Unlimited Resources • Unlimited waste	• Quasi-cyclic Material Flows • Use of Energy and Limited Resources • Limited waste	• Cyclic Material Flows • Reuse and recycling of resources • Zero waste

Fig. 3 Ecosystem types [20]

resources and eco systems. It has been predicted that the next Industrial Revolution will be toward the current scarcity and many business organizations have started emulating the new business model—Natural Capitalism [16]. The infrastructure of Natural Capitalism is environmental quality. Environmental quality is based on technology that unites economics and ecology, innovations and entrepreneurship and interdependence between eco systems and economy. The futuristic view shows a shift in economy with emphasis on enhancement of resource productivity. Resource productivity uses resources efficiently, lowers pollution and also provides meaningful jobs to people across the globe. Improvement in quality and flow of life supporting services enhances human welfare promoting economic and environmental sustainability.

- **Blue economy systems**: The concept highlights the integration and interdependence of different production systems by using the resource available in cascading systems—the waste of one production system becomes the input for the next system. Blue Economy is an open-source report which contains informative case studies which provides solutions to different problems based on the characteristics of the local environment and using gravity as the source of energy. The manifesto of the Blue Economy report is '100 innovations for 100 million jobs in the next 10 years' [21]. Many organizations are given awards for reducing pollution levels by change in industrial technology or processes similar to a thief being awarded for robbing smaller amounts of money. Temporary solutions are not sufficient and there is a need to look for other means to solve day to day problems. One resource that is very helpful to mankind and that has been tested for millions of years is 'Mother Nature'—a treasure chest of transformational remedies and solutions to problems. The logic behind nature can become the foundation for bringing about many changes in concept, design, structure and products/processes.

Water crisis and extreme weather are great challenges that face humans. In a location which receives half-inch rain a year, the Namib desert beetle wing structure is the best model for the development of the water collecting system which has the capacity of harvesting water from the fogs that are part of the strong winds that blow across the desert [22]. A surface film design was developed from the beetle wing which had bumps on the wings to collect water and valleys with scales that could collect the water like a funnel that was thinner than human hair [23, 24]. A film was developed to recover 10% of the water lost and around 50,000 water cooling towers are erected annually and each system retrieves around 500 million liters of water per day. Similarly, the dimensional spinning techniques developed by the 'golden silk orb weaver' spider helped in the manufacture of silk tubes and filaments for implantable medical devices and also for the substitution of Titanium in airplane parts and razors [25–27]. Coffee and agro waste have been used for growing mushrooms, while the spent mushroom substrate is utilized as high-quality animal feed. This shows how the waste becomes a resource for cascading systems. Similarly, the nano scale pacemaker devised from the whale's regulating current, the dragon-fly technique for solar power generation, chelation techniques of bacteria to separate metals and the concept of acidity in the stomach + alkalinity in the small intestine used for clean environment in building interiors. The sustainability of the planet is dependent on Nature's Mastery of Brilliant Adaptations [MBA] [28] to help the earth to evolve with solutions that are pollution free and ecologically safe.

The problems faced by the Linear Economy have developed new technologies which have ultimately emerged as the Circular Economy. The extractive model has been transformed to a regenerative representation with a focus on economic, environmental and societal benefits, new designs with recycling, reuse and zero waste. The textile industry has grown by leaps and bounds and has gone through various levels of productivity and economic development. However, the impact of this growth is multifold and the footprints on ecology and environment is varied. A thorough study of the textile industry systems and procedures should be undertaken to incorporate changes and assess the results of the growth in terms of sustainability. The textile industry has majorly contributed to the economy through foreign exchange earnings and employment generation. Hence the impact of the industry should be analyzed to understand the differences in linear and circular economy and its pace toward circular economy.

2 Global Textile Industry and Its Growth

Globalization has highlighted the dependence of the production and financial assemblies of countries with cross border transactions resulting in the international division of labor. The wealth of a country relies on economic agents in other countries or in other words there in economic integration on a global scale. Global economy has transnational firms and financials operating in different world markets and is not restricted by national level boundaries, policies, politics and economic constraints. Apart from cross border interactions, capital mobility has a more complex and deep influence on global economy [29–31] along with progress in technology and fierce competition. All these theories show that there is greater openness and liberalization of domestic and foreign trades far beyond the influence of domestic policy makers and state managed economic activity. The new forces for economic development in a global economy are open and free markets, transnational connections and corporations and new information technologies. These forces have great influence on the policies followed in many developing nations.

Globalization has changed the phase of industrial production from the developed countries to the developing countries called the newly industrializing economies of the Third world. East Asia has become the hub of all types of productivity leading to high per capita growth rate, domestic savings and investment, higher education and reduction in income inequality. The period 1965–1990 saw the remarkable growth of eight economies in East Asia namely Hong Kong, Korea, Singapore and Taiwan called the 'Four Tigers'; Japan; China; Thailand and the new industrializing economies (NIEs)—Southeast Asia, Indonesia and Malaysia. These eight economies are considered as high performing Asian economies (HPAEs). During 2015, the four

Asian tigers have surpassed Japan (GDP) and rank second to China in economic success [32].

In the early 1960s, after the Second World War and the Korean War [1950–53], there were major advances in air travel and telecommunications being advantageous to all the four countries which had well-established ports, strong infrastructure, advanced trade economies educated population. The policy of these countries supported heavy investment and infrastructure building in industry, luring foreign investments with tax benefits and providing compulsory education for its young population thereby guaranteeing a good work force.

In the early 1960s, the global economy was just starting to recover after the traumas of the Second World War and the Korean War of 1950–1953. Tentative world peace combined with major advances in air travel and telecommunications meant that borders were opening up around the world, and the four 'tigers' were perfectly positioned to benefit. All four countries boasted long-established ports and developed trade economies, highly educated populations, as well as robust post-colonial infrastructure (as a result of the British influence in Hong Kong and Singapore, the Chinese in Taiwan, and the Americans in South Korea).

The four 'tiger' governments took this opportunity to invest heavily in industrialization, building major industrial estates, offering tax incentives to foreign investors, and implementing compulsory education for its young population in order to secure the future of the workforce. Soon, these countries were in high demand, exporting everything from textiles and toys, to plastics and personal technology. The specializations of the 'Four Tigers' are given in Fig. 4. The strong economic growth of these four countries helped them to withstand 1997 Asian Financial Crises and the 2008 Global Financial Crisis [33]. These four countries are commonly featured in the IMF list of the most prosperous and stable economies of the world.

There are many factors that helped in the growth of the textile industry. In the case of the 'four tigers' the main reason for the rapid growth was high rates of private investment (exceeding 20% of GDP) coupled with rising human capital due to primary and secondary education. Sustained growth could be attributed to equal income distributions and high levels of productivity; in turn productivity increase is due to the allocation of capital to high yielding investments and combining technology with the industrial economy. The public policies like stable macro-economic environment, consistent legal framework and absence of price controls and other

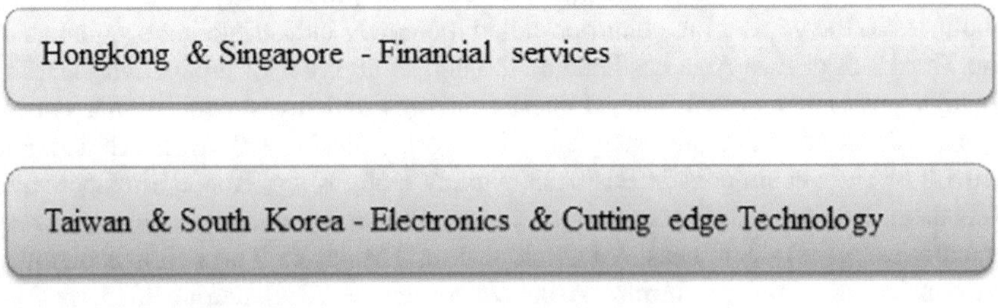

Hongkong & Singapore - Financial services

Taiwan & South Korea - Electronics & Cutting edge Technology

Fig. 4 Specializations of four tigers

distortionary duties served to boost growth. In general, after the First World War the global textile industry grew to a better status providing opportunities for improvement and advancement [34].

- **Textile Industry of Georgia**
  Georgia is a country at the intersection of Europe and Asia and former Soviet Republic. The textile industry has a significance influence on the inhabitants of the state as Georgia became the leading producer of textiles in the South. Silk production was started initially but after the invention of the cotton gin, cotton manufacturing became the primary industry. Small mills emerged and the number rose from 19 in 1840 to 38 in 1850; 1880's and 1890's saw major expansions with the help of steam power technology and in 1899 the Georgia Institute of Technology started the Textile Department to provide formal education to young students. The 1900's saw the emergence of the production of carpets, bedspreads and the tire fabric for the automobile industry [35].
  The World War I (1917–1918) marked a great demand for uniforms for the American troops and many mills were given government contract. Racial segregation, Civil war and the boll weevil affected the growth of the textile industry. World War II saw the use of nylon and production of technical textiles like camouflage nets, life rafts, gas masks and uniforms. Modernization and automation transformed the industry and in 1994 the North American Free Trade Agreement (NAFTA) eliminated quotas and saw the emergence of the low-cost Asian textile companies [36].
- **Indian Textile Industry**
  The Indian Ocean societies have viewed textiles as a main article of trade for local needs and exports. Based on inscriptional evidence during 1000–1500 AD, there were around 132 weaving centers in South India [37]. During the 18th century, many mechanical inventions around 70 in number like spinning frame, steam engine, spinning jenny, cotton gin and the Crompton mule spinning, witnessed the Industrial revolution. The opening of the Suez Canal in 1869 and the development of shipping and railways attracted foreign competition leading to the decline of the ancient industries. In 1854, a cotton mill was established in Bombay by Mr. C. N. Davar [38].
  In the 17th century, the dominance of the East India Company saw the dumping of the Manchester fabrics to be sold in India and Indian goods were sold in British markets. The decade 1860–1870 was not favorable to the cotton textile industry as cotton prices rose high due to civil war in the US. The British started mills to compete with the Indian owned mills. A report in 1936 stated that 1400 million yards of khadhi was produced by the two million handlooms in India which contributed to 26% of the total quantity consumed in India [39]. Rapid growth in cotton spinning was witnessed in the 70's, 80's and mid 90's after which the industry was down by the import tariffs implemented by the Government.
  During the post war period after World War I, there was a boom due to raising the exports to China. During this period, Bombay became the center of cotton mills and provided employment to 60% of employed population in the country. During

World War II, though handlooms faced problems, 1943–1943 the production of fabric was 818 million yards when compared to 177 million yards during 1938–1939. After independence, there were a lot of changes and the industry with technological improvements in fibers, machinery and the entrance of synthetics which saw a tenfold growth from 1.5% to 15% during 1965 & 1975. During the 20th century, the powerlooms were introduced in the weaving industry. In 1942, there were 15,000 powerlooms which increased to 22 lakh powerlooms by 2011–2012. Many structural changes and policies brought into existence the decentralized small scale powerlooms which proved to be greater in cost effectiveness.

- **The textile industry of China**

  Late 1800's cotton yarn was imported from British India to help the weavers of Northern China. The benefits for operating in China were many namely huge market, raw materials and cheap labor, low costs for transport and longer working hours. Foreign enterprises were established by the British & Japanese apart from the Chinese owned ones and after the Sino Japanese war (1904–1905) they enjoyed tariff exemptions and lesser political restrictions.

  During World War I, manufacturing of goods came to a standstill and the textile industry of both China and Japan developed rapidly to fulfill the needs of the world. The machine tools for the textile industry, which was initially imported, was produced and in 1920's Japan became the main supplier of machines, followed by China [40]. The mills in China grew rapidly, but one third of the cotton yarn and half of the cotton goods produced were controlled by Japanese firms and these mills withstood the depression of the 1930's [41].The Japanese took advantage of the utilities infrastructure provided by the Germans and the knowledge of machinery development by the local people to establish seven modern spinning mills with increased productivity due to electric powered weaving machinery [41]. In 1937, when the Sino Japanese war broke out, the Chinese held 2.75 million spindles & 25,500 looms, while the Japanese owned 2.38 million spindles and 33,800 looms.

  The devastation that occurred during the war was made up after the defeat of Japan in 1945 by China which helped China to own many of the well-equipped and profit earning industries [42]. After Japan's retreat, the Chinese Government took over the mills and merged them into the Chinese Textile Industries Corporation to help them to revive quickly and produce textiles for the army uniforms. When the Chinese Communist Party rose to power all private owned textile operations shifted to Hong Kong.

  With these stages of development the textile industry in China has earned a trade volume that has increased 27.11 times during the past 25 years (from the Reform & Opening). In 2005, textile exports moved up by 15.4% of China's total exports and 24.4% of the world's total exports in textiles [43, 44]. China also holds the world level first position in the production and exports of textile products and clothing with the gross clothing exports accounting for one fourth of the total global clothing trade. The Chinese clothing industry comprises of the clothing processing and the fashion industry. In the recent years, the Chinese apparel industry has surpassed the GDP and textile industry [45].

These stats have changed considerably and the textile and clothing industry which was focused on exports is focusing on the domestic market and slowly losing its momentum and price competitiveness in the US market. According to the 12th Five Year Plan, China will concentrate on industry upgradation, building a green industry, moving its production from the east coast to the west for upliftment of the local economy and to expand domestic consumption. Chinese industry is shifting to more value-added products like technical textiles and will continue to improve in skill and capabilities [45, 46].

Currently, China is the world's largest manufacturer and exporter of textiles and apparels as the output volume of this industry occupies more than 50% of the world, the capacity for processing and export of the fiber is the highest in the world and the international market share is more than one third of the globe. These credentials have showcased the production value of the textile industry as 7% of the GDP of China; the exports and the employment status highlight that this industry provides numerous opportunities for China. The prediction is that China will establish and maintain a certain advantage in the world in the next three to five years by having an impeccable industry chain and an extensive domestic market [47].

- **Textile Industry of England**

During the seventeenth century, the imports of Indian cotton fabrics through the East India Company, created many opportunities in terms of design, material and patterns. In Lanchashire, the first cotton industry was established between 1660–1770 and between 1770 and 1830 technological developments helped Lanchashire to have a competitive edge in the international markets and after 1830 Lanchashire also competed with the Indian market [48].

England was dependent on manual labor to support their cottage industries like weaving, in the 1700s. In 1733, the loom with the flying shuttle was invented by John Kay enabling England to produce more fabric is lesser time. Mechanization of the manual work was possible with the invention of Richard Arkwright water frame for cotton spinning and Spinning Jenny by James Hargreaves. This was, followed by the invention of the Spinning Mule by Samuel Crompton, which was a combination of the Spinning Jenny and the Water Frame [49, 50]. In 1785, Edmund Cartwright patented the power loom helping England to dominate the textile industry of the world [51]. In 1840, the Brunswick Mill was established in Manchester, as a midcentury model mill, with fireproof internal construction and powered by a double beam engine. By 1850, the development was vast and the mill housed 276 carding machines, 77,000 mule spindles, 20 drawing frames, 50 slubbing frames and 81 roving frames [51–54]. In 1994, this building was listed under the Grade II listed buildings which pronounced it as a building of special interest and takes every effort to preserve them under the listed buildings of England and Wales [55].

Currently, the findings of the Alliance Project Report states that the Textile and clothing industry in UK is worth around GBP 9 billion (US $ 13.71 billion) to the national economy and supports almost 90,000 to 100,000 jobs; the UK also has significant capabilities in spinning, weaving, knitting, apparel and in technical textiles, materials and composites [56]. The 'National NBrown Textiles Growth Program', 2013, is a textile grant growth program backed by the online catalog and stores retailer N Brown Group, UK. Since its inception the program has granted funds to the tune of GBP 9 m to 94 companies and will be using another GBP 30 m of private sector investment. Supported by Department of Business Innovation & Skills (BIS), Mark & Spencers and others, this program has been involved in the creation of 1,600 jobs and 115 apprenticeships in England during the first year [57–59].

- **American Textile Industry**

  In earlier times, the American textile industry was an outbreak against British enforcement and taxes on goods from England. People boycotted British products and started producing their own local products under the name 'home spun' which started with fabric, apparel and spread to the provision of uniforms and blankets for the soldiers of the Continental Army of the Revolutionary War [60]. Many technological inventions in England paved the way for faster production of textiles and the British maintained the inventions as secrets under the threat of treason. Despite all these hurdles, in 1789 Samuel Slater used his designing skills and experience in England to replicate the construction of textile mills in New England, USA, to initiate the industrial revolution in America. Samuel Slater will always be remembered as 'Founder of the American Industrial Revolution' [61, 62].

  The first textile factory was established in 1814 in Lowell, Massachusetts and this industry became a forerunner for all other industries to come. Soon Massachusetts became the center for textiles and economic growth. During World War I and II, many textile industries catered to the needs of the war and USA started following the principle of self-sufficiency. The post war period saw economic growth and sustenance of the industries during economic crisis. Late 1990's textile manufacture transformed to textile marketing as the industries were shifted overseas due to low-cost labor and materials [63]. North Carolina and South Carolina, which were the major hubs for textile manufacturing, have recently caught the interest of the Chinese companies. China has identified that labor, materials and energy in Carolina is lucrative and has turned its focus to this state. The resurgence of the textile industry and economy is in the near horizon for the state of California. The US Bureau of Labor Statistics estimates that the textile industry in the US, in 2015, employs 579,300 people and contributes to around $100 billion to the GDP of the nation. The exports include fiber, yarns, fabrics, made-ups and apparel totaling to $21.6 billion and occupying the fourth largest exporter in the world [48, 64]. In 2017, the National Council of Textile Organizations (NATO) has reported that the American Textile Industry has reached in 2016, $74.4 billion in terms of shipments of textiles and apparels which accounts to an increase of 11% since 2009. The exports of fiber, textiles and apparel by the US are $26.3% billion in 2016 and the capital expenditure for textile and apparel production amounts to $ 2 billion in 2015, it has also been estimated that the employment in the US textile supply chain is around 565,000 in 2016 [65].

  The growth of the textile industry from the earlier days to the current times show how the industry has crossed many stages in terms of output, export, specialization, technology, economy and upliftment. Many factors have favored the growth of the industry, but in all the cases studied this industry has nurtured the livelihood of many families and societies. The statistics and analysis undertaken are very many to understand the reasons for growth like favorable environment for the growth of raw material, cheap labor, extensive market, technological improvements, human capital and many more. Leaving this aside, we should also study the impact of the growth and development of the textile industry across the globe. The current status of the textile industry in all the different parts of the world is well-established and serving the economy of the nation, but what about the impact on the society and the environment which are the other pillars of sustainability?

3 Issues with Linear Models

The population on the earth is growing by leaps and bounds and soon there will be overcrowding as the space and natural resources remain the same. When we look at the pictures of our natural resources say water ways, taken earlier, they seem to be bountiful and flowing with lush green cover and mountains all blending to give a holistic amalgamation of everything natural. But today the same places which were beautiful and bountiful seem to have lost their charm as man has over utilized the available natural resources without any concern for replenishment. The current world population is 7,612,097,061 as per the worldometer clock taken today [66–68]. The forecast by the United Nations Report for the growth of the population is that it would reach 8 billion by 2025, 9 billion by 2040, 9.6 billion by 2050 and 11.2 billion by 2100. The estimate of the growth of the world economy is 26 times of this century putting enormous pressure on the resources which is being used at 160% [69]. This alarming rate speaks of the atrocities carried out on the use of the freely available natural resources without any concern for the consumption of our future generations.

The world economy has been on the 'linear' mode for a very long time and all the development during the Industrial Revolution, World War I & II and post war periods was achievable with the help of massive use of the resources from nature. The motto of linear economy is 'take-make-dispose' or 'take-make-use-dispose' [70]. In this economy, products are manufactured with the use of natural resources and after they are used they are thrown away as waste. The production flow is continuous and creation of value is by the maximization of the products produced and sold. Waste was enormous both during manufacture as well as after disposal. Soon scarcity of resources was evident and the 'Reuse Economy' was initiated. Reuse is the secondary use of a new product for a conventional reuse or to fulfill a different function. Reuse is different from recycling as reuse is found in down cycling processes and recycling is carried out in upcycling processes, recycling is the breaking down of used items to identify raw materials for the manufacture of new products. E.g., Reuse—concrete is shredded and reused as road filament; recycling—concrete grounded into grains and used as raw material for construction [71].

The third option is the circular economy where the design of the product is planned to keep the raw material stream pure throughout the value chain so that it is a onetime investment, easy to retrieve and can be used multiple times for a certain function [72]. Waste is reduced and if raw material is required it is taken from nature sustainably so that there is no damage to environment and humans. The sustainability of linear economy is improved by '*eco efficiency*' where more goods and services are produced with lesser resources and minimizing the waste and pollution. Further, the negative impact is reduced to help to postpone the moment at which our system is overburdened. In the case of circular economy '*eco-effectivity*' is the norm where the system minimizes the negative impact by creating maximum benefit from process change, system change and innovations thereby reducing entropy [73, 74]. Figure 5 shows the characteristics of the different types of economies.

Design development plays a major role in highlighting which economy the product will be classified. Apple has developed glued iPad of 9.4-mm thickness which requires the company to replace the battery on the basis of payment; Google on the other hand has offered the Nexus 7 which is 10.4-mm thick, but allows for repair and extends the life of the device [77]. Apple has optimized its design to appeal to the sleek customer sacrificing repair, reuse and remanufacturing while Google has augmented for an audience who is engineering minded and tinker with the device. The device is compatible with the circular economy as it is possible for reuse, remanufacture and recover. The difference in the design is one millimeter, but the difference it creates in the mindset of the consumer is enormous and the label it carries for sustainability will call out loud to the customers and complete the sale.

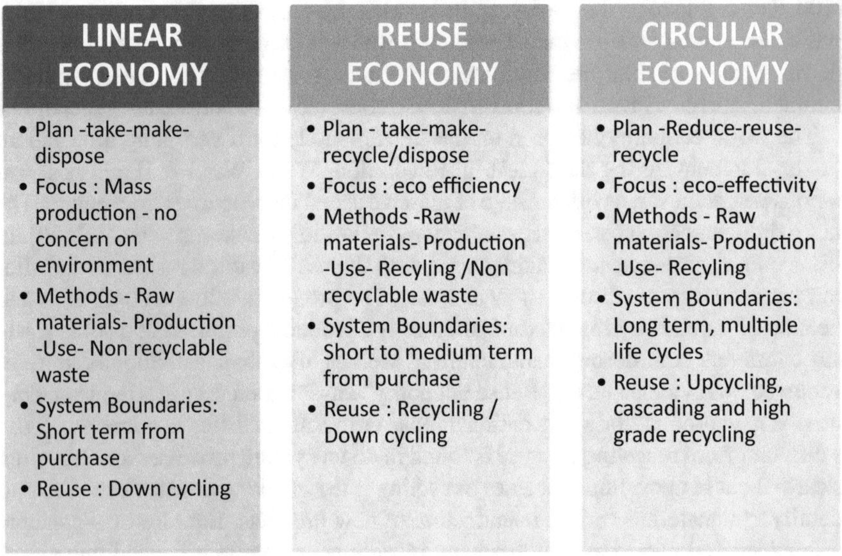

LINEAR ECONOMY	REUSE ECONOMY	CIRCULAR ECONOMY
• Plan -take-make-dispose • Focus : Mass production - no concern on environment • Methods - Raw materials- Production -Use- Non recyclable waste • System Boundaries: Short term from purchase • Reuse : Down cycling	• Plan - take-make-recycle/dispose • Focus : eco efficiency • Methods -Raw materials- Production -Use- Recyling /Non recyclable waste • System Boundaries: Short to medium term from purchase • Reuse : Recycling / Down cycling	• Plan -Reduce-reuse-recycle • Focus : eco-effectivity • Methods - Raw materials- Production -Use- Recyling • System Boundaries: Long term, multiple life cycles • Reuse : Upcycling, cascading and high grade recycling

Fig. 5 Characteristics of linear, reuse & circular economies [75, 76]

3.1 Impact of the Linear Economy Model

- **Risks in supply**
  The linear economy is demanding in terms of material and energy. The period from 1900 to 2000, was a golden era which witnessed a GDP rise by twenty times and unknown levels of materials richness and consumption. The linear economy policy was to extract materials from nature for production and consumption without any concern or plans for dynamic regeneration to continue the work of Earth. The period also observed a situation where there was abundant availability of consumer goods with high quality and low cost which was made possible by technological developments in production, global supply chain and lower labor inputs. The transition of the low-productive circular economy to the linear economy was seen in the developing nations. The international supply chains looked upon the customer as the King of Consumption and aimed at satisfying his needs. The Organization for Economic Co-operation and Development (OCED) reports that the annual consumption of a civilian is 800 kg of food & beverages, 120 kg of packaging and 20 kg of new clothing and shoes and 80% of these culminate in incinerators, waste water or landfills [78].
  Uncertainty is common in the linear economy as the planet has fixed amount of materials and the availability of these materials depends on replenishment mechanisms. Products and processes in many industries are dependent on materials and geographical developments and their limited availability has caused demand and higher prices reducing the profit margins. Agricultural productivity is slowing down, and there is a decline in soil fertility and nutritive value of foods; the risks for food safety and security are rising. Though the processes have been optimized and standardized there is still room for efficiency increase and the gains are not good enough to show a marked advantage or differentiation.
- **Volatility in Prices**
  Challenges in the supply have brought and era of 'resource stress' where unpredictable prices of materials will suffice, environmental degradation will rule, shortages and disruptions in supply will be common and political/military control over resources leading to major shifts in the natural environment, link between resource systems and distribution of income and power on a global scale. All these challenges will change the assumptions on availability of resources, their extraction, production, processing and consumption and also on the power/nation that will have control over the global resources leading to conflicts and presenting lesser chances for sharing of resources. Since 2006, the prices of commodities have been fluctuating causing increase in prices and giving rise to risks in investments in material supply. When prices become volatile the chance of accessing resources become distant and this may lead to crisis and war.
- **Materials of Importance**
  Many industries like metal industry, electronics and electrical industry and the computer industry use critical materials for their production. Raw materials include metallic minerals, industrial minerals, construction minerals, wood rubber

etc. Some of the materials are traded under stock exchanges, while the others are not. Base materials like aluminum, copper, lead, nickel, tin, zinc are traded in stock exchanges, e.g., London Metal Exchange (LME) [79], while others like cobalt, gallium, indium or rare earths are traded as per the supplier. The major rise in demand of raw material was seen during 2002–2008 due to economic growth, which was marked with unprecedented high prices [79]. This can create situations like difficulty to access materials due to stock accumulation, price hedging for future contracts and long term contracts. Dependence on critical materials will be impacted by price fluctuations, availability and one cannot make predictions for future trends and analysis.

- **Global Connectivity**
  Global trading has brought about interconnectedness of products. Oil rich countries trade oil for food leading to linkages. Fuel and water are important requirements for the manufacture of products and scarcity of raw materials will impact the availability and prices of products. Regional price changes can reach the global market due to integration of financial markets and good transportation system. During times of need all markets tend to integrate into the global value chains and systems.
- **Increase in Demand for Material**
  The world's population will soon be reaching 8 billion by 2025 coupled with the growth in prosperity as the GDP is on the rise. This causes a high degree of consumption calling for more material and energy to manufacture products. It has been estimated that if the world consumed resources at half the rate of the US consumption then 'strategic materials' (copper, tin, silver, chromium zinc) will be depleted by 40 years. Some rare earth metals if continued to be consumed at the existing rate will be depleted within a decade [80]. The estimate for resource consumption increased by two fold during 1980–2020 and by 2050 the consumption would become three fold. Further, the extraction of resources is on the rise and the supply is closer to exhaustion. The demand for resources has also put additional pressure on environment in terms of locating new sources for raw material extraction and its allied impact. These issues will cause resource scarcity and there will be sharp increases in material costs. In a survey in 2012, conducted by the Manufacturers' Organization the estimate states that 80% of the CEOs' of manufacturing companies indicated that raw material shortage was a high risk to their business [81].
- **Degradation of Eco systems**
  The linear economy has manufactured innumerable goods with no concern for the environment. Pre-consumer waste during manufacture of goods and products and post-consumer waste after use of the product by the consumer has generated large quantities of materials that are not used, but may be incinerated or dumped in landfills. This has caused over burdening of the eco system that is not able to perform its usual function of providing food, processing of nutrients and offer building materials and shelter. It has been reported that the annual production of plastics in 2013 is around 78 million tons which has 90% virgin feedstock; after use 14% goes for incineration, 40% to landfills, 32% as leakage and 14%

collected for recycling. Of the 14% collected for recycling 4% is process loss, 8% cascaded recycling and only 2% goes back into the manufacturing loop [82]. The Sustainable Europe Research Institute (SERI) estimates that in the OCED countries around 21 million tons of materials enter the waste stream and are not physically incorporated into the final product each year, e.g., Parting materials in mining, dredged materials from construction [78]. Recycling of materials is in the pipeline, but may not be lucrative as the global end-of-life recycling rates (EOL-RR), the recycled content (RC) and the old scrap ratio (OSR) are very low leading to inherent limitations in the recycling processes. Of 60 metals used for the study, only 18 metals had an EOL-RR above 50% (3 metals above 50% and the remaining metals between the ranges 25–50%) [83]. Further, it should also been noted that these primary metals are abundantly available at very low cost. This estimate will help recycling organizations to improve recycling methods and estimate the viability of the process. Though the process may be expensive research is required as these metals may become scarce in the years to come.

- **Decreased product lifetime**
  Consumers today are over enthusiastic looking at advertisements and promotional messages. They are updated with the present day availability of products which have many extra features compared to the older ones. This has resulted in shorter lifespan of products and a tendency to keep changing products to stay updated with the latest trend. The useful life of a product has become shorter as specified—a study estimates that the median lifespan of products has reduced between the years 2000 and 2006 by 3% for washing machines, 5% laptops & PCs, 9% hot water and coffee, 11% printing and imaging equipment and 20% for small consumer electronics and accessories. This craving for new products by consumers will make manufacturers design products that has a shorter life span and the need for quality products which serve longer periods will diminish. This situation will have a bearing for procurement of raw material for manufacture of new products and also result in higher percentage of waste.
- **Need for accountability**
  Environment, climate change and sustainability are the three major focus centers in any industrial production and economy. Laws and regulations are constantly changing and every industry whether manufacturing or retailing is accountable for the waste it produces in any stage of the product life cycle. A study conducted in six Indian cities showed that 20% of the waste was recovered by 80,000 rag pickers in recycling three million tons. Every ton of waste that was recycled earned INR 24,500/- and the emission reduction was 721 kg of carbon dioxide per annum [84]. In India, the accountability lands on the government to segregate waste and to clean the environment but the law has started placing the onus on the manufacturing and retailing sectors or the consumers who have made use of the product. Waste audits are essential to control waste during manufacture and use by sustainable design and development.
  Consumers, politicians and the government are concerned about the environmental depletion and the impact on the human race. A study performed in the US on the consumer's accountability for the carbon emitted, the participants view

carbon footprint label and it has an impact on their purchase. A carbon footprint label displays the CO_2 emissions that is related to the manufacture of a product (production, transportation, use and disposal). This shows the consumer the choice of the purchase and their responsibility during purchase. Similar studies have been conducted which highlight the role of the consumer his accountability toward the environment in terms of choice and use of products [85, 86]. When consumers refrain from unsustainable products the consumer's choice is based on the ecological footprint which can hail/detain the power of a brand. The linear economy will soon show the negative side as no concern is given to the environment. Policy makers and nations across the world will look forward to the circular economy and sustainable business.

Growth and development at the cost of the planet has shown the negative side and linear economy has lost its charm. Growth coupled with sustenance of resources and materials and biodiversity is the next phase. This calls for consideration before squandering resources to fulfill the needs of the human race. The population is growing rapidly and human needs are ever increasing- a waste free growth model is the requirement and the linear economy needs to be replaced by the circular economy where preservation and regeneration will produce rewards. The time has come to invest in developing a circular economy that will help to build and nurture the environment for our future generations.

4 Textile Waste and Closing the Loop

The insatiable demand for clothing has led to the immense growth of the textile industry to be in the band wagon of large industries of the world. It is a big contributor to several economies around the world and also termed as the second largest polluting industry. The processes involved for the manufacture of the basic unit—the fiber to the last stage—the product/fashion is very many and pollution results all along the tract of production, retailing and consumption. Different waste streams in all the three states of matter—gas, liquid and solid converge to air pollution, water pollution and solid waste pollution. The Chinese Environmental Statistical Year book, 2010, has stated that Chinese Textile Industry has produced two and half billion tons of waste water [87]. The chemical content of the effluent includes toxic organic auxiliaries, anions, salts, metals, complex metallic compounds, biocides and surfactants lead to foaming, coloring and aquatic toxicity.

4.1 Waste in the Textile Industry

Air pollution is another problem in textile manufacture and can occur as diffusive sources (pollution due to solvents, spills & waste water treatment) and point sources (pollution from ovens, boilers and storage tanks). Fibers whether oil based or natural are consumers of large quantities of energy and chemicals. It is estimated that about one million tons of chemical dyes are used every year [88, 89]. Foul smelling gases like Sulfur and Nitrogen from boilers, hydrocarbons and other volatile compounds from drying ovens and curing tanks are some of the reasons for air pollution. The solid waste pollution from textile manufacturing is used clothing and textile manufacturing waste ending in landfills. Some of the contaminants are manufacturing waste (fiber waste, packaging waste, remnants from fiber and fabric processing, sludge etc.). These land up in landfills and release methane and may land up in water bodies killing marine life.

The consumption of textiles in the global scale in 2015 was 95.6 million tons [90] and the share of each class of fibers is given in Fig. 6. Fiber production calls for about 2 trillion gallons of water and 145 million tons of coal which is just one of the energy sources used in the textile industry. Coal mining is the primary source of pollution, while coal burning causes the secondary form of pollution. Surface mines change the ambience of the place due to the use of explosions to get coal and the dust and water runoff effect environment. Underground mines are a source of methane gas. In 2015, methane gas from coal mining and abandoned coal mines in the US, accounted for 10% of the US gas emissions and 1% of total US GHG emissions [91]. Apart from this coal burning emissions include SO_2, NO_x, CO_2, particulates, mercury and other heavy metals, fly ash and bottom ash causing environmental concerns like smog, acid/colored rains, toxic environment and human concerns like respiratory disorders, cardiovascular and cerebrovascular effects. These are only the pollution effects of using coal. What about the other fossil fuels that are used for textile manufacturing?

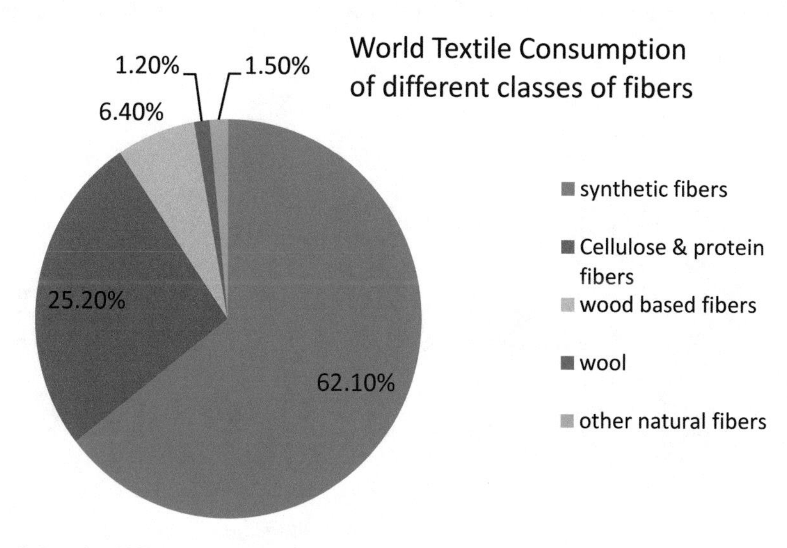

Fig. 6 Lenzing [90]

4.2 Waste in the Clothing Industry

Fashion predominates the use of apparels. Consumers are fashion conscious and want to upkeep with the latest trends. This leads to fast fashion and most of the clothes purchased is disposed within one year. This linear system has a great impact on the natural resources and ecosystems. Figure 7 shows the greenhouse gases in each phase of clothing manufacture and use. Further, a study states that the number of times a garment is worn by a consumer has reduced by 36% on a global scale when compared to the number of times worn 15 years ago, while in China the utilization of garments has reduced by 70%. Clothes as throwaways cost USD 460 billion per year. 60% of Germans and Chinese consumers own more clothes than required [92–94].

Some alarming statistics are being presented here to understand the extent to which the textile industry has polluted the environment. The greenhouse gas (GHG) emissions from textiles in 2015 are around 1.2 billion tons of carbon dioxide equivalents. It has been estimated that the total emissions in 2015 is 21 times more than the total emissions if all international flights and naval ships [95]. It has also been reported that 20% of water pollution by industries is due to textile industries discharging water from dyeing and finishing of textiles. Synthetic textile like polyester, nylon and acrylic shed small broken microfibers during the washing process. The wash water end in the ocean and the estimate says that around half million tons of these fibers [96] reach the sea annually which affect the fauna and flora and in turn land in the esophagus of animals and humans when consuming sea food. A report states that the microfiber dump in oceans would reach 22 million tons by 2050 and the carbon foot print would touch 26% from a 2% in 2015 as seen in Fig. 8 [97, 98]. The need

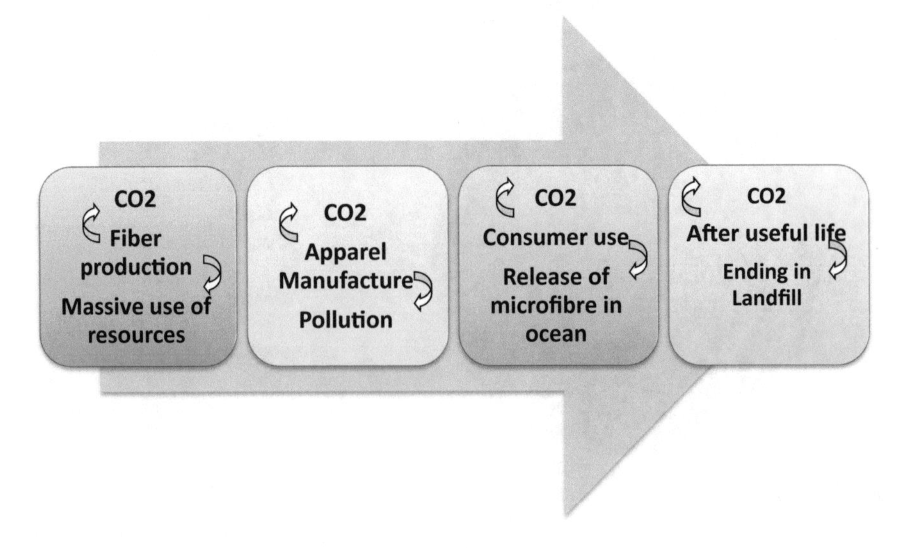

Fig. 7 Negative impacts of clothing manufacture [92]

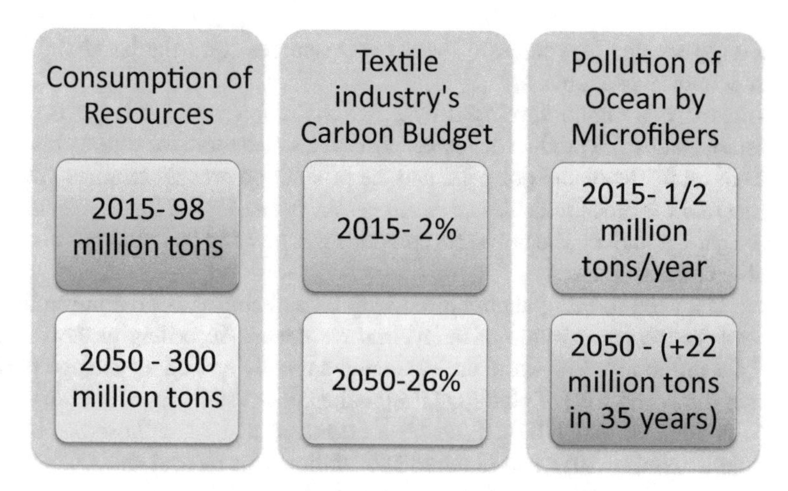

Fig. 8 Prediction of the negative impact of the textile industry 2050 [95]

to convert the linearity of the economy to a circular one is evident and this calls for immediate attention and action.

4.3 Closing the Loop

Any product which is sold in the market goes through all the stages from manufacture to use and decline as stated in the apparel product life cycle [99]. Once the design hits the market, the manufacturers want to get the best out of the season and utilize the demand for the product by massive production. But after the use, the stage of decline is the most problematic area as many consumers want the product to be out of their domain and go on to the new next trendy product that is in fashion. These products end up in the landfill and take time to decompose, this process leads to the emission of methane a greenhouse gas.

Reuse and recycling help to use the entire raw material or product after functional use in a closed loop production system. Waste out of one production cycle is converted as resource material for the subsequent production cycle resulting in a non-ending flow in a circular economy. A process is classified as 'zero waste' when some parts the product is suitable for disassembly and reuse and the other components are converted by recycling for the next production cycle. Water is the best example, available in nature in many forms to suits different needs, but there is no down cycling or degrading of water molecules in any form. Two material flow types [100] have been identified in a circular economy namely resources that land back after use to Nature termed as 'Biological Resources' and those resources that continuously circulate into the production system without landing in the biosphere termed as 'Technical Resources'. These two resources form the backbone of the circular economy phenomena.

'Climatex' is a textile developed by a Swiss Company which is the first fully compostable industrial product developed in 1995. A substitute for cotton, this fabric is made of ramie, wool and polyester and the production process requires 30% less water; the fabric is biodegradable and the processing waste is used for the manufacture of felt or garden mulch. The polyester component is low and hence comes under the negligible category [101].

There is a need to bring about closed loop manufacturing for the prevention of waste and for the preservation of the natural resources. According to the USEPA, about 14.3 million tons of waste are generated annually and on an average the US consumer throws 65 LBS of clothing [102] in the UK the statistics of clothing waste in 2016 is 300,000 tons [103]. It has been estimated that 82 million tons of fiber was manufactured in 2011 and required 145 million tons of coal and a few trillion gallons of water to process the fiber and manufacture the product. Added to this is the consumer's part of buying unnecessary products and throwing them before their functional use is over (Fig. 9).

The I: Collect [I: CO] is a global solution provider for recycling collecting old clothes and shoes from almost 60 countries. Sorting and reuse or sorting and upcycling are the methods used for closed loop recycling, while recycling helps in providing raw material (insulation composites for automobile and the construction industry) for various industries through the open recycling method. Retailers like H &M, Levi's, Puma, Forever 21 & North Face some of the participants of I: CO's sustainable movement [104].

The average woman in the US discards clothing within 5 weeks of purchase and 85% of the clothes purchased are thrown away. Of the $1770 annual shopping budget, about $1445 worth of clothing is replaced every year. #30 WearsChallenge advocates

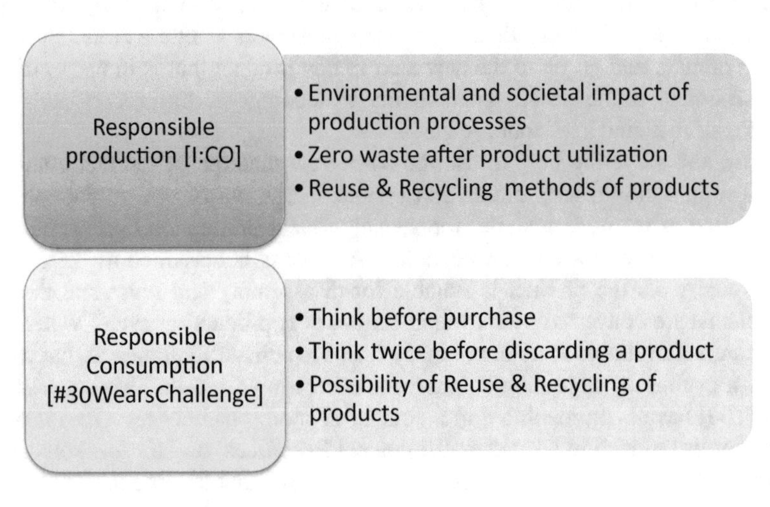

Fig. 9 Responsible production & consumption

consumers to wear a piece of clothing atleast for 7 months or 30 times before it is discarded. This situation is due to the speed of fashion change and the focus is on design and not on quality. Slowing down of fashion is ethical and also calls for more usage of garments. This is also economical as a shirt worth $10 when worn 5 times, the cost per wear is $2, but if it is worn 30 times the worth per wear ($1.17) is better. This thought process shows the responsibility of the consumer in terms of purchase, use and economics. When a consumer thinks that a particular item in his wardrobe is not required, one should see if it could help someone or it could be recycled. Currently, there are many centers in each country that take back used clothes for passing it on to a second hand user or sending it to a recycling agency.

4.4 The Reach Forclosing the Textile/Fashion Loop

H&M, one of the world leaders in apparel business and recording artist M.I.A., have taken efforts to close the fashion loop into a circular economy. The 'World Recycle Week' scheduled between April 18–24th, aims to collect two million pounds of unwanted clothes form 3600 stores across the globe. Since 2013, consumers who return back garments of any brand will receive a 10% off in their next purchase and during the World Recycle Week the discount is 30%. The effort of the 'Garment Collection Program' on the part of H&M has collected 25,000 tons of clothing and presented its first 'Close the loop' clothing line in 2014. Apart from this, 38 million t-shirts worth of material was donated for reuse, while recycled polyester from 40 million plastic bottles [105, 106].

Inditex, the Spanish multinational clothing group and Australian textile manufacturer Lenzing have joined hands to identify opportunities to recycle post-consumer textile waste. The environmental strategy includes the starting of a pilot project in 2016 for offering clothing collection service on a free basis with the assistance of a charity Caritas and a logistics company Seur. Under this service, customers can donate old clothes to the delivery person, while receiving the products from new orders. The collected clothes will be sent to Lenzing for conversion into new fiber, the target being 500 tons for the current year and a long term goal of 3,000 Tons [107] over the following years. The recycled TENCEL fiber will become a raw material for industries who have a sustainability core in their manufacture.

Transparency in manufacture and reporting transparency is a marketing tool for many fashion retailers to increase customer base. A number of industry workers have started a global movement to track transparency in manufacture and retailing. They have launched the Fashion Transparency Index announcing the top 100 fashion companies based on their disclosures of their social and environmental impact. H &M is ranked in the highest percentile somewhere between 41 and 50% [108].

Similarly, there are other retailers marching forward to close the loop to safeguard the environment—Common Thread Initiative & The Cleanest Line by Patagonia; Fabric of Change Initiative by Asoka & C & A Foundation, India; Texcycle project by Coop & Tell-Tex group; Textile Waste Diversion, Canada; Worn Again, UK [108–110]. Closing the loop to bring about a circular economy may be easy to say, but it holds a lot of challenges and presents boundless opportunities.

5 Challenges and Opportunities

Circular economy aims at having materials in the flow of production so that there is no waste. Zero waste is the target, but achieving closer to that is the first step when compared to the linear production systems. Different strategies are adopted by many manufacturers and retailers like reuse as seconds, recycling by conversion to raw materials or secondary raw materials, renting clothes, completely biodegradable or partially degradable products, buy back of new clothes for return of old ones. These approaches are important in achieving a circular economy. However, to make the production cycle continuous without waste serious attention ought to be given to certain parameter or criteria to enable the manufacturer to achieve a closed loop process.

5.1 Requirements for Closed Loop Manufacturing

Circular economy is possible if certain criteria have been met. Planning ahead is required otherwise it may not be possible to undertake services like repair refurbish or renew products. Sustainable product design is an important parameter. The initial phase of designing should facilitate disassembly of the components to enable repair or replacement of the broken parts; at the end of life of product with due concern for recycling. Further, the design should assist in recycling the technical parts or can be biodegraded/composted to biological nutrients. Certification of nonuse of hazardous chemicals that ensure safe recycling, biodegrading technique or composting. Any critical material used, if any, should have a separate traceable loop. With regard to collection, consumers must be willing to donate their unused garments to the collection points with or without rewards or subsidies. Effective collaborations between consumers, charity organizations, manufacturers, retailers, government and society to collect the waste material, repair and refurbish them to create new products or to extend the life of products.

5.2 Challenges

The global supply chain is very extensive and products manufactured have different places of origin. The challenge here is that when large scale biodegraded material results how and where will it be stocked, disposed or used again. Used apparels have many individual components and trims which are not textile in nature, e.g., Labels, leather patches, zippers, buttons etc. Special efforts are required for separating them and making them fit to be reused or broken up into their raw materials by recycling. Apart from this recycling of cotton fiber is more difficult than polyester as fiber breakage is common. Many of the brands also sell blended fabrics and recycling them is difficult.

High value/textile to textile recycling is the target for the future. Any leftover textile fabrics or old garments would have to go back into the textile manufacturing process as raw material to create a circular manufacture. When this stage is reached the use of virgin raw material would be drastically cut down thereby reducing the negative impacts on environment. However, there are technical challenges that have to be addressed. Not all styles can be created with recycled yarns and the suitability to the product design is of utmost importance. Recycled yarns have lower strength and certain finishes and resin treatments cannot be given. Around 95% of the recycled yarns are used as weft yarns as the warp yarns bear the stresses of weaving. The raw material, though sorted, used for recycling is from different sources and each batch may produce different types of yarns which will generate different outcomes. Further, most of the processes in recycling is manual, e.g., collection, sorting but this may become laborious when the number of garments increase. Optimization of the blend ratio of recycled content of fibers in yarn and yarn in fabric requires many trials. Wearer trials have to be undertaken before taking it for commercialization. In most cases, multicolored yarn results and can be used effectively for designing apparels. When the recycled fiber content increases higher percentage of down cycled products results. Virgin materials are added to make it look closer to products made from virgin materials.

Raw material cost is an important part of production and reports state that the use of fossil fuel has increased by a factor of 12 and the extraction of material resources has increased by 34 times. Further in the European Union, a citizen uses 16 tons of materials/year of which 6 tons are wasted and 3 tons go to landfills [111]. Economics are agents which lure manufacturers to take up a stream of production or manufacture. The challenge is the cost of recycling in terms of infrastructure, personnel and capital. Recycling may create material lower in quality than virgin materials. The era of cheap and bountiful supply of virgin raw materials has come to an end. With this background, a series of questions arise which have to be addressed. Virgin raw materials are scarce and expensive, how will we cut down the cost of a product in the near future which has only a very small percentage of recycled material, how many times can we use recycled materials in our process as the quality of the recycled materials will deteriorate; after the required/optimized number of use what will be the fate of the recycled materials and how will it impact the quality of the new products and environment; what will be the compromise in quality that will be accepted by the buyer with regard to recycled products; how will the complaints from the customer be addressed if the quality parameters are not up to the mark, while putting the recycled garment to use; when virgin raw materials are expensive recycled fibers will also be expensive based on demand; recycled material certification for quality and environment is required. The World Business Council for Sustainable Development has estimated that by 2050 there must be 4 to 10 times resource efficiency for real economic development and this trend should be experienced by 2020 [111].

One of the main objectives of the OCED Environmental strategy for the First Decade of the twenty-first century is decoupling growth [112]. When the growth rate of the environmental pressure is lesser than the economic driving force (GDP) over a given period of time decoupling occurs. OCED has specified 16 indicators for decoupling environmental issues from economic ones. They include climate change, air pollution, water quality, waste disposal, material use and natural resources. Another 15 indicators focus on the economic side in four areas like energy, transport, agriculture and manufacturing. Sustainability in product design, management of environmental resources, reuse, recycling, substitution of materials for resource savings all contribute to decoupling growth. This calls for transformation and change in manufacturing where innovation and resource efficiency is the key factor.

5.3 *Opportunities*

To enable closing the loop in production, technology and innovation need to go hand in hand and new methods and processes are to be perceived to convert polyester or cotton to recycled yarn, fabric and apparel without loss in quality. If this process is continuous, there will be no need to use natural resources, e.g., oil for the manufacture of virgin raw material. Oatshoes, Netherlands have brought into the market 100% biodegradable nontoxic shoes and leather bags that can be used as containers for plants [113]. Macdonald, UK and Worn again have set up a collection center for their staff uniforms and enabled closing of the loop for the uniforms and using biodiesel from used cooking oil for travel of their fleet [114].

The reuse of garments and textiles will eventually land up in landfills. Recycling of textiles give rise to secondary low-value products which can be given higher status under the closed loop recycling. A good example for this concept is the new generation swim wear manufactured with 100% recycled nylon from waste streams like used fishing nets and carpets. Econyl, X2 plus, Returnity and SaXcell, [115] manufactured by the company Aquafil, are the brand names of the swim wear collection that have multiple life cycles which has brought appreciation from reputed brands like Koru Swimwear and Adidas. Similarly, 'Returnity' is a100% recyclable polyester that is used for Dutch work wear with take back arrangements of old ones. In both the examples cited this concept is to be extended to other items of clothing from other types of waste.

Chemical recycling helps to give rise to a better grade fibers when compared to mechanical recycling. Some of the technologies used to recycle polyester are given here. 'Carbios' is a process of using enzymes to Biorecycle PET by depolymerization to monomers; Gr3n uses microwave radiation to hasten the depolymerization of PET for recycling; Loop Industries: recycling of PET waste into branded resins; Resinate Material group: the process of converting PET waste to high-performance polyester polyols using glycolysis for the formation of oligomers to manufacture raw material for various applications like lubricants, solvents, resin coating, personal and home care, artificial leather and thermoplastics [116–118] Worn Again: derived a dissolution process to separate the constituent fibers from PET/Cotton blended fabrics.

Manual sorting of garment according to their material content is the current system. When garment volumes for sorting become larger this operation may be difficult. Fibersort technology has come to the aid of recyclers which automatically sorts large volumes of garments and products by identification of the fiber composition. This mechanism can detect both pure fiber and blends up to 14 fiber compositions with a target productivity of one piece per second as the scanning of the product takes a few milliseconds [119–121]. This technology will help the recycling institutions to cycle resources continuously through the supply chain and also facilitate high value recycling. The sorting machines also help to procure data on related areas like yields, sales value, productivity of human operators and help to optimize the supply chain.

It has also been proposed to issue loyalty cards to enthuse the consumers to identify what they actually possess and want to recycle. 'Polymers in your closet' or 'Cotton Credits' will help to lease products through their fiber content rather than the garment [122]. Further, rewards and subsidies can be offered to the customers based on what they return to or have the lowest returns to the system.

5.4 The True Potential of High Value Textile Recycling—Case Studies

- **Closed loop Blankets from Dutch Work wear**
  In USA and UK around 75–85% of used textiles and in Netherlands, around 30 Kt/year uniforms are sent to landfills for incineration. Uniforms from reputed agencies and government organizations prefer it to be incinerated to safeguard brand reputation and misuse of uniforms. Such waste could be identified for fiber extraction either by mechanical or chemical methods. Uniforms, from the Dutch Defense with a composition of 50:50 cotton/polyester on an average, were first sorted for color and type by BigaGroep and cut into smaller pieces and sent to Recover, Spain. The cut garment pieces were cleaned to remove the non-recyclable parts (buttons and zippers) and then sent for mechanical extraction of the fibers with the help of steel wires. 3/1 Nm recycled yarn produced had a composition of 80% recycled fiber and 20% virgin polyester fiber and they were used as weft yarns for the conversion to blankets for humanitarian aid. The economics for the recycled yarn developed are given in Table 1. The final Recover recycled yarn had a blend composition of 40:40:20 = recycled cotton fiber/recycled PET fiber/virgin PET fiber. The environmental impacts assessed highlighted that water consumption and energy consumption reduced by 87% (Virgin 26–34.5m^2 Recycled) and 42% (Virgin 115–67 GJ Recycled), respectively, due to the absence of dyeing and the

Table 1 Recover recycled Yarn economics from Dutch Defense Workwear [123]

Description	Weight (kg)	Description	Weight (kg)	Description	Waste %
Dutch defense workwear	1.45	Cleaned defense workwear	0.95	Cleaning waste	(0.51) 35.17%
Cleaned defense workwear	0.95	Recycled fibers from defense workwear	0.90	Recycling waste	(0.05) 5.26%
Recycled fibers from defense workwear	0.90	Virgin polyester fiber	0.22	Spinning waste	(0.12) 10.71%
Final recycled Yarn from defense workwear	1				

greenhouse gas emissions reduced by 33% (Virgin 4.66–3.13 tons Recycled) in comparison with the use of 'all virgin polyester' [123] (Table 2).

- **Closed Loop Denim from G-Star RAW denim**
  The collaboration of G-Star RAW with Wireland Textiles and Recover in 2016 gave rise to a lifecycle assessment study for the development of closed loop recycled denim. Recycled denim fabric from returned denim garments has 12.5% premium in price when compared to their virgin equivalents. The returned denim inventory was handled by G Star in three different approaches. In the first approach (Incineration), the used denim was incinerated to produce electricity which was

Table 2 Recycled Denim fabric economics from returned denim [123]

Description	Weight (kg)	Description	Weight (kg)	Description	Waste %
Denim garment waste	1.00	Cleaned denim materials	0.65	Cleaning waste	(0.35) 35%
Cleaned denim materials	0.65	Recycled fibers from cleaned denim materials	0.62	Recycling waste	(0.03) 5%
Recycled fibers from cleaned denim materials	0.62	Virgin cotton fiber	1.45	Spinning waste	(0.207) 10%
Recycled cotton weft yarn (30% RC) Ne 9/1	1.86	Virgin cotton warp yarn Ne 8/1	2.79	NA	0
Final recycled denim fabric (12% RC) (7.65 m)	4.65				

RC—Recycled Content

fed back to the energy grid at the expense of paying for transportation and incineration of the waste. The impact was assessed with the alternative production of grid electricity. In the second approach (down cycling) where the used denim is sorted and cut for extraction of fibers to produce nonwovens at the cost for transportation, sorting and bundling, conversion to nonwoven and auditing of the process. The impact was estimated with similar products manufactured with virgin fiber. The third approach (High value recycling). Transports the used denim to Wireland Textiles where it is sorted and bundled; then, it reaches recover for cleaning and recycling and spinning into yarn with 30% recycled fiber which contributes to 40% of fabric weight; it is then converted to fabric with 12% recycled content at the weaving mill. The denim fabric weighs 0.66 kg per meter and is 162 m wide. G star pays for the transport and for the purchase of the recycled denim fabric. The impact of this approach is valued in comparison with the virgin equivalent denim fabric. The environmental impacts measured state that water consumption reduced by 9.8% (9.017–81.210 m^3), not much difference in energy consumption 4.2% (455–436 GJ) and carbon emissions by 3.8% (33.8–32.5 tons). The environmental impact study for all the three approaches were also given as in Table 3.

- **Closed Loop Recover yarns from post-consumer waste**
  Post-consumer used garments are sorted by Dutch charity Sympany as rewearable and recyclables. The recyclables are further sorted for color—white, denim and multicolored. Recover, Spain separates the non-recyclable parts like zippers and buttons called cleaning process. The remaining textiles are cut and converted to recycled fiber followed by carding, spinning and plying into 20/2 Nm for the production of new fabrics. Recover uses 'Color Blend' process thereby avoiding 80% of the dyeing process. These recycled yarns are converted to woven or knitted fabrics for product development.

Three different colored yarns and one white cream yarn have been produced by this high value recycling process. Recycled material 1—Light Bruma/yellow/gold yarn is composed of 70% post-consumer textiles (87.5% cotton +10.2% PET + 2.3% other fibers); Recycled material 2—Anthracite/gray yarn is composed of 30% recycled post-consumer rPET from bottles; Recycled material 3—Multicolored yarn is composed of 61.25% recycled cotton +7.14% recycled PET from textile waste +30% rPET from bottles +1.54% other post-consumer fibers. The environmental impact of the white cream yarn was estimated. The water consumption reduced by 62% (286–108 m^3) as water intensive dyeing process is averted; energy consumption reduced by 33% (88.7–59 GJ) as post-consumer waste avoids virgin raw material consumption like PET which is energy intensive and dyeing is not required; greenhouse gas emissions reduced by 18% (3.84–3.16 tons) due to the use of post-consumer waste instead of virgin material (Table 4).

Table 3 Environmental Impact of three approaches undertaken [123]

<table>
<tr><th rowspan="2">Sl No</th><th rowspan="2">Parameters analysed</th><th colspan="3">Approach 1
Incineration vs Grid electricity</th><th colspan="3">Approach 2
Down cycling vs virgin fiber</th><th colspan="4">Approach 3
High Value Recycling vs virgin fabric</th></tr>
<tr><th>m3</th><th>Liters</th><th>Showers</th><th>m3</th><th>Million Liters</th><th>Showers</th><th colspan="2">m3</th><th>Million Liters</th><th>Showers</th></tr>
<tr><td>1</td><td>Net water savings</td><td>10</td><td>10,000</td><td>± 210</td><td>2.698</td><td>2.7</td><td>± 57,000</td><td colspan="2">8.837</td><td>8.8</td><td>± 186,000</td></tr>
<tr><td rowspan="2">2</td><td rowspan="2">Net energy savings</td><td></td><td colspan="2">Equivalent</td><td>GJ</td><td colspan="2">Equivalent</td><td>GJ</td><td colspan="3">Equivalent</td></tr>
<tr><td>0</td><td colspan="2">NA</td><td>10</td><td colspan="2">Half the annual energy of 1 household in UK</td><td>19</td><td colspan="3">The annual energy of 1.9 households in UK</td></tr>
<tr><td rowspan="2">3</td><td rowspan="2">Net impact in $C0_2$ emissions</td><td>ton</td><td colspan="2">Equivalent</td><td>Ton</td><td colspan="2">Equivalent</td><td>ton</td><td colspan="3">Equivalent</td></tr>
<tr><td>-1</td><td colspan="2">± two long distance flight per person (negative impact)</td><td>0</td><td colspan="2">NA</td><td>+ 1.3</td><td colspan="3">± two long distance flight per person (savings)</td></tr>
</table>

Table 4 Recycled Yarn economics from post-consumer waste garments [124]

Description	Weight (kg)	Description	Weight (kg)	Description	Waste %
Post-consumer textile waste	1.05	Cleaned P–C textile waste	0.87	Cleaning waste	(0.18) 17.14%
Cleaned P–C textile waste	0.87	Recycled fibers from P–C textile waste	0.78	Recycling waste	(0.09) 10.34%
Recycled fibers from P–C textile waste	0.78	Recycled polyester fiber (rPET)	0.34	Spinning waste	(0.12) 10.71%
Final recycled yarn from P–C textile waste	1				

5.5 *Roadmap for a Circular Economy*

The roadmap is planned in two segments as in Table 5. One segment deals with the sustainability measures in the production cycle [raw material to product development-distribution to disposal] and the second segment looks into the steps, systems and strategies to be taken by the stake holders to move toward the circular economy. Whatever be the suggestions given, new technologies and research has to take upper hand to make concepts to happen in real time situations. Laboratory results are to be carried to a commercial scale with the cooperative and collaborative effort of all the stake holders. Assessments, evaluation and reports must be made transparent easily accessible without any payment as these will give rise to new ideas and field level and market level implementation. These efforts will focus the need to achieve a circular economy and help nature to replenish resources for the future generations.

6 Conclusion

The circular economy in textiles has a repetitive lifecycle with design as its first phase- a design representing durability, long life with reuse or recycling facility. To move to a circular economy a huge transformation is required with a strong collaboration with all the members of the supply chain. Education, research, finance and regulation- all require a mindset toward a circular economy where material input and output is of primary importance. Transformation can be acquired by changing the consumption and preferences of consumers, public, industry and authorities. Sharing, renting or buying of products as a service help in increasing the utilization with economic viability. An example for this concept is seen in designer handbags being rented during the weekend, wedding gowns and lehangas [125], are rented in India for marriages and other important festivities.

Table 5 Sustainability measures towards a circular economy

Fiber

- Limited use of virgin fiber
- Protection of natural resource
- Material and process planning will reduce energy usage
- Use of side streams can be taken by means of environment impact assessment & environmental permit processes
- Maximize the length of material life cycle and opportunities of recycle and reuse
- Agricultural production with cleaner techniques and lower emissions and use of bio processes
- Cascading use of resources - to increase productivity
- and to make efficient use of scarce and valuable raw materials
- Materials retain their potential for reuse
- Use of healthy materials that have been grown or made under conditions healthy for the environment and humans

Manufacturing

- Use of recycled fiber (mechanical & chemical recycling) for yarn manufacture
- Use of recycled yarn in the weft direction
- Efforts and research for developing recycled yarns in the warp direction
- Waste reduction at source
- Coloured yarn from waste does not require dyeing
- In case the yarn needs dyeing only a small percentage is added to give richness to the base colour

Design

- Eco design – design for disassembly, recycling
- Avoid single use materials eg. Single use of plastic
- selection of materials suitable for recycle and reuse
- Sustainability models: Design for material recovery
- Dismantling design-Design for disassembly
- Design of long life garments
- Post consumer waste – upcycling for remanufacturing

Product development

- Analysis of waste for type of product development
- Down cycling – breaking a product to component materials and reusing the materials for low value products
- Down cycling to be taken in the initial stage of product development
- Upcycling – reuse of discarded material/objects to create a product of higher quality/value when compared to the original product
- The aim of product development is high value recycling

Retailing

- Will sell more services instead of goods
- inform customers about maintenance and repair services, environmental impacts, materials and future use in the final phase of the lifecycle of the product
- Distribution- transport coordinated between different sectors; renewable fuels; transfer of products can be done with jointly owned transport; optimized and clean transport

Consumption & Disposal

- products should not be disposed but the option for reuse to be analyzed
- full use of product with service, part change and repair as necessary
- after full use the product should be dismantled to be reused in the lifecycle of other products

ZERO WASTE – DOWN CYCLING – UP CYCLING – HIGH VALUE RECYCLING – MATERIALS BACK TO PRODUCTION STREAM

⇩

CIRCULAR ECONOMY

POLICIES & STRATEGIES FOR ZERO WASTE – DOWN CYCLING – UP CYCLING – HIGH VALUE RECYCLING – MATERIALS BACK TO PRODUCTION STREAM

- Slow fashion concept to replace fast fashion
- Demands for sustainable products
- Purchase of goods during need
- Durability and good quality to be of criteria for selection of products
- Look out for eco friendly and

- make eco design compulsory - get design approved for its eco features
- setting of centers for eco design - for research and assistance to manufacturers and entrepreneurs
- promote zero waste awareness
- subsidies and incentives for zero waste
- Extended Producer Responsibility for products with waste and throw away policy at end of life cycle
- penalty on use of certain unrecyclable products eg. Polystyrene in packaging

- provide parts and components for repair and reuse
- provide maintenance services for the products they sell
- Reduce the level of fixed and single use parts
- Tie ups with other industries for use of by products and waste materials - Industrial Symbiosis
- Material passport-Better

(continued)

Table 5 (continued)

Consumer	Government	Manufacturers/Service Agents/Distributors
recycled products • Refurbish/repair/reuse of apparel to ensure longevity of garments for use • Return products at end of life for reuse& recycling • Have concern for environment • Encourage other customers to join for a sustainable cause	• bonus for use of recyclable raw materials • Tax increase for waste incineration and land filling to promote recycling • Loans and financial aids to organizations based on EPR and reduced VAT • Setting up of a Zero waste center for research and implementation of substituted processes • Advisory board for circular economy - continuous development of requirement standards for circular economy • Promote research, development, testing, demonstration and market development for circular solutions and technologies • footprints of products to be submitted to the government yearly like income tax • Services for such work to be made freely available for the business community to fulfil the task of submitting the environmental and social report to the government	information & transparency to the consumer on content and potential for recirculation as product • Environmental and social report of the organization made available to public • Horizontal collaboration for circular economy -supplier to manufacturer & investor to leader

SITRA, Finland has been nominated for the world's premier circular economy award organized by the World Economic Forum and Young Global Leaders in 2017. The roadmap for the circular economy has been published which highlights the projects and policy recommendations for bringing about a circular economy. The aim of the roadmap is to make Finland a leading circular economy in the world by 2025 with 250 ideas, 60 projects and many pilot studies. Industrial excess heat will be used for production, shopping could be done in circular economy shopping centers, traveling in driverless buses, driving cars on biogas, and repurposing offices for living spaces. Solutions for the climate problem have been spelt out in their publication 'Nordic Green to Scale' to reduce emissions to meet the targets set by the Paris agreement [126].

Refibra™ is a reborn fiber made by Lenzing is the most sustainable fiber from natural raw materials. The use of bioenergy and closed loop production of almost 99.7% has helped the TENCEL fibers to be awarded the EU prize for the most ecofriendly fiber. Refibra uses the same technology of the TENCEL fiber and also uses waste cotton scraps and textiles [127] as raw materials for the next cycle of production thereby closing the loop and avoiding textiles in landfills. A new identification system makes it possible to identify the Refibra fiber in the finished textiles highlighting the transparency in the supply chain. These fibers serve as brand ambassadors for the manufacturer and the brand is licensed after the textile is certified.

Many such cases can be given to show that the world is moving towards a circular economy with many new and innovative mechanisms in the agenda toward sustainability. Zero waste is the mainframe of the circular economy and the adoption of research findings and new business models will move this attempt to new heights. Transformation and transitions are important for both fundamental and day-to-day technical issues and these systems should be embedded in all the activities of business proposals and transactions. Assessments and evaluation reports will help to highlight the direction and impact of the activity or change incorporated; publication of the reports and findings will serve as guides to other members of the business society.

Stakeholders would have to extend their support at all levels and help to create the shift—moving from the linear to a circular economy. The benefit will reduce the use of raw materials and help the environment to replenish and indulge in the growth of all things in nature to serve the needs of mankind till times immemorial in the context of the future.

To illustrate the negative impacts of industrialization, here are a few quotes…..

> 'Humans are the only creatures on Earth that will cut down a tree, turn it into paper, then write 'save the trees' on it'—Quotling.com
>
> 'Modern technology owes ecology an apology'—Director of Green Earth Affairs, Zimbabwe.
>
> 'We don't wanna live in a trash can, stop making it one'.
>
> 'Climate change is no longer some far-off problem; it is happening here, it is happening now'—Barack Obama
>
> To transition to a circular economy requires using renewable energies. And the most renewable are the human energies of empathy, honesty and integrity—Rob Peters.

References

1. Purt J (2012) Talkpoint: rethinking the global economic system. https://www.theguardian.com/sustainable-business/rethinking-global-economic-system-sustainable-capitalism. Accessed 2 March 2021
2. Malley K, Singh R, Duan T (2020) 2nd Law of thermodynamics. https://chem.libretexts.org/@go/page/1923. Accessed 9 March 2021
3. Brein Het Groene (2021) How is a circular economy different from a linear economy. https://kenniskaarten.hetgroenebrein.nl/en/knowledge-map-circular-economy/difference-circular-linear-economy/. Accessed 9 March 2021
4. Pearce DW, Turner RK (1990) Economics of natural resources & the environment. Johns Hopkins University Press, Baltimore, USA
5. Oberfield C (2018) Anaerobic-digestion-vs-composting. https://renergy.com/anaerobic-digestion-vs-composting/. Accessed 9 March 2021
6. Chynowetha DP, Owens JM, Legrand R (2001) Renewable methane from anaerobic digestion of biomass. Renew Energy 22:1–8
7. Manavalan E, Jayakrishna K (2019) An Analysis on Sustainable Supply Chain for Circular Economy. Procedia Manuf. 33:477–484
8. Jawahir IS, Dillon OW, RouchKunal KE, Joshi KJ, Venkatachalam A, Jaafar IH (2006) total life-cycle considerations in product design for sustainability: a framework for comprehensive evaluation. In: 10th International research/expert conference proceedings—trends in the development of machinery and associated technology, Spain
9. Stahel WR (2010) Performance Economy. St Martin's Press, New York
10. McDonough W. (2021) Cradle to cradle. http://www.mcdonough.com/cradle-to-cradle/. Accessed 9 March 2021
11. Hamilton T (2008) Whale inspired wind turbines. https://www.technologyreview.com/2008/03/06/221447/whale-inspired-wind-turbines/. Accessed 9 March 2021
12. Biomimicry Instiute (2008) Biomimicry—innovation inspired by nature. https://s3-us-west-2.amazonaws.com/oww-files-public/c/c1/IBE_-_biomimicry_lecture.pdf. Accessed 9 March 2021

13. Whalepower Corporation 2021. The science. https://whalepowercorp.wordpress.com/the-science/. Accessed 9 March 2021
14. Benyus J (2006) Biomimicry—innovation inspired by nature. https://www.aspenideas.org/sites/default/files/transcripts/Transcript_Biomimicry.pdf. Accessed 10 March 2021
15. Arnarson PO (2011) Biomimicry. http://olafurandri.com/nyti/papers2011/Biomimicry%20-%20P%C3%A9tur%20%C3%96rn%20Arnarson.pdf. Accessed 10 March 2021
16. Anonymous (2021) Natural capitalism-the next industrial revolution—session eight—ENVI 5050. www.unicamp.br/fea/ortega/extensao/20-NaturalCapitalism.ppt. Accessed 10 March 2021
17. Manahan SE (2019) Environmental science and technology—a sustainable approach to green science and technology, CRC Press, UK. https://www.routledge.com/Environmental-Science-and-Technology-A-Sustainable-Approach-to-Green-Science/Manahan/p/book/9780367390129
18. Cliff R, Druckman A (2016) Taking stock of industrial ecology, Springer, Cham, UK. https://www.springer.com/gp/book/9783319205700s
19. Ayres RU, Ayres LW (2002) A handbook of industrial ecology, EE Publishing, Cheltenham, UK. https://www.elgaronline.com/. Accessed 10 March 2021
20. Lifset R, Graedel TE (2021) Industrial ecology: goals & definitions. http://planet.botany.uwc.ac.za/nisl/ESS/Documents/Industrial_Ecology_Overview.pdf. Accessed 10 March 2021
21. Ellen Macarthur Foundation (2017) Schools of thought—circular economy. https://www.ellenmacarthurfoundation.org/circular-economy/schools-of-thought/blue-economy. Accessed 10 March 2021
22. Ask Nature Team (2017) Water vapor harvesting- Darkling Beetles. https://asknature.org/strategy/water-vaporharvesting/#.Wp4n_ehubIU. Accessed 10 March 2021
23. Randall I (2014) New device pulls water from thin air. https://www.sciencemag.org/news/2014/06/new-device-pulls-water-thin-air. Accessed 10 March 2021
24. Rain harvest Company (2014) Water vapour harvesting by the namib desert beetle. www.rainharvest.co.za/2014/07/water-vapour-harvesting-by-the-namib-desert-beetle. Accessed 10 March 2021
25. Tokareva O, Jacobsen M, Buehler M, Wong J, Kaplan DL (2013) Structure–function-property-design interplay in biopolymers: spider silk. https://www.ncbi.nlm.nih.gov/pmc/articles/PMC3926901/https://doi.org/10.1016/j.actbio.2013.08.020 Accessed 10 March 2021
26. Römer L, Scheibel T (2018) The elaborate structure of spider silk Prion 2(4):154–161. https://doi.org/10.4161/pri.2.4.7490
27. Saravanan D (2006) Spider silk—structure, properties, spinning. JTATM 5(1):1–20. https://wet.kuleuven.be/wetenschapinbreedbeeld/lesmateriaal_biologie/saravanan.pdf. Accessed 10 March 2021
28. Balance3 (2014) The blue economy—applying nature's MBA. http://balance3.com.au/the-blue-economy-applying-natures-mba/. Accessed 11 March 2021
29. Bairoch P and Kozul-Wright R (1996) Globalization myths: some historical reflections on integration, industrialization and growth in the world economy. http://unctad.org/en/docs/dp_113.en.pdf. Accessed 11 March 2021
30. Gordon D (1988) The global economy: new edifice or crumbling foundations? https://newleftreview.org/I/168/david-gordon-the-global-economy-new-edifice-or-crumbling-foundations. Accessed 11 March 2021
31. The Changing World (2021) Globalisation. http://developmentandglobalisation.weebly.com/globalisation.html. Accessed 11 March 2021
32. Gaw K (2016)The story behind the four Asian tigers. https://www.idealsvdr.com/blog/the-four-asian-tigers/. Accessed 11 March 2021
33. Velocity Global (2017) Why the four Asian tigers should be on your rad. https://velocityglobal.com/blog/why-four-asian-tigers-should-be-on-your-radar/. Accessed 11 March 2021
34. Preston LT (1993) The East Asian miracle—economic growth & public policy. http://documents.worldbank.org/curated/en/975081468244550798/pdf/multi-page.pdf. Accessed 11 March 2021

35. Williams A (2007) Textile industry. https://www.georgiaencyclopedia.org/articles/business-economy/textile-industry. Accessed 12 March 2021
36. Andrews MG (1987) The men and the mills: a history of the southern textile industry. Mercer Univ Press, Macon, GA
37. Ramaswamy V (1983) Textiles and weavers in medieval South India. Oxford University Press, Bombay, p 7
38. Sastry NSR (1947). A statistical study of Indian industrial development, Thacker & co ltd, Bombay, pp 1–2
39. Dantwala ML, Vakil CN (eds) (1937). Marketing of raw cotton in India, Longmans, Green and Co. Ltd, Calcutta, p 9
40. Broggi CB (2016) Trade and technology networks in the Chinese textile industry—opening up before the reform, Palgrave Macmillan, US
41. Kubo T (2004) chinese cotton industry in the 20th century. In: Proceedings of the GEHN conference 5th conference. http://www.lse.ac.uk/Economic-History/Assets/Documents/Research/GEHN/GEHNConferences/conf5/KuboGEHN5.pdf. Accessed 12 March 2021
42. MacDonald S, Pan S, Somwaru A, Tuan F (2004) China's role in world cotton and textile markets. In: 7th annual conference on global economic analysis, Washington, DC, pp 17–19. http://citeseerx.ist.psu.edu/viewdoc/download?doi=10.1.1.498.3657&rep=rep1&type=pdf. Accessed 13 March 2021
43. Anderson K (1992) The new silk roads, Cambridge University Press, number 9780521392785. https://ideas.repec.org/b/cup/cbooks/9780521392785.html. Accessed 13 March 2021
44. Yuan T, Xu F (2021) China's textile industry international competitive advantage and policy suggestion. https://www.bpastudies.org/bpastudies/article/view/24/53. Accessed 13 March 2021
45. Irun B (2017) The textile and apparel market in China. http://ccilc.pt/wp-content/uploads/2017/07/eu_sme_centre_report_tamarket_in_china_2017.pdf Accessed 13 March 2021
46. Textile Today (2017) China is building tech intensive textile industry, leaving low value business. https://textiletoday.com.bd/china-building-tech-intensive-textile-industry-leaving-low-value-business/. Accessed 14 March 2021
47. Business wire Inc. (2016) China textile industry overview 2017–2021—research and markets. https://www.businesswire.com/news/home/20161003006057/en/China-Textile-Industry-Overview-2017-2021---Research-and-Markets. Accessed 13 March 2021
48. Broadberry S and Gupta B (2005) Cotton textiles and the great divergence: Lancashire, India and shifting competitive advantage, 1600–1850. http://www.iisg.nl/hpw/papers/broadberry-gupta.pdf. Accessed 14 March 2021
49. Sea.ca (2003) The industrial revolution -innovations of the industrial revolution. https://industrialrevolution.sea.ca/innovations.html. Accessed 14 March 2021
50. Skuola network (2021) Industrial revolution & innovation. https://www.skuola.net/letteratura-inglese-1800-1900/industrial-revolution-innovations.html. Accessed 14 March 2021
51. Mount B (2010) American textile history. http://www.brahmsmount.com/blog/american-textile-history/. Accessed 14 March 2021
52. Historic England (2021) Brunswick Mill. https://historicengland.org.uk/listing/the-list/list-entry/1197807. Accessed 14 March 2021
53. Grace's Guide Limited (2021) Brunswick Mill. https://www.gracesguide.co.uk/Brunswick_Mill. Accessed 14 March 2021
54. Paarkinson-Bailey JJ (2000) Manchester—an architectural history. Manchester University Press, UK
55. Gov.uk (2018) Principles of selection for listing buildings. https://assets.publishing.service.gov.uk/government/uploads/system/uploads/attachment_data/file/757054/Revised_Principles_of_Selection_2018.pdf. Accessed 14 March 2021
56. Woodard R (2015) UK textile & clothing industry poised for growth. https://www.just-style.com/analysis/uk-textile-and-clothing-industry-poised-for-growth_id124343.aspx. Accessed 14 March 2021

57. Innovation in Textiles (2015) Industry talk-investment in UK textile manufacturing to create 20,000 jobs by 2020. https://www.innovationintextiles.com/investment-in-uk-textile-manufacturing-to-create-20000-jobs-by-2020/. Accessed 14 March 2021
58. Wightman-Stone D (2015)Textiles industry awarded 19.5 million pound grant. https://fashionunited.uk/news/fashion/textiles-industry-awarded-19-5-million-pound-grant/2015021515493. Accessed 14 March 2021
59. Newgate (2015) Can fashion sourcing from the UK become a reality? https://www.newgatecomms.com/blog/index.php/2015/03/13/can-fashion-sourcing-from-the-uk-become-a-reality/. Accessed 14 March 2021
60. Loeschen D (2019) The evolution of American textiles. http://blog.mixerdirect.com/blog/the-evolution-of-american-textiles/. Accessed 14 March 2021
61. Bellis M (2019)A history of the textile revolution. https://www.thoughtco.com/textile-revolution-britains-role-1991935. Accessed 14 March 2021
62. Bowen A (2016) The American textile industry was woven from intellectual espionage. https://www.atlasobscura.com/articles/the-american-textile-industry-was-woven-from-intellectual-espionage. Accessed 14 March 2021
63. Howell L D (1964). The American textile industry. https://naldc.nal.usda.gov/download/CAT87201751/PDF. Accessed 14 March 2021
64. Hanichak G (2021) American textile industry. https://study.com/academy/lesson/american-textile-industry.html. Accessed 15 March 2021
65. NCTO (2017) State of the US textile industry. http://www.ncto.org/2017-state-of-the-u-s-textile-industry/. Accessed 14 March 2021
66. Worldometer (2021) current world population. http://www.worldometers.info/world-population/. Accessed 15 March 2021
67. Live population (2018) Live world population clock. https://www.livepopulation.com/. Accessed 15 March 2021
68. UNFPA (2020) World population trends. https://www.unfpa.org/world-population-trends. Accessed 14 March 2021
69. The World Counts (2021)The world population is growing by over 200,000 people a day. http://www.theworldcounts.com/counters/shocking_environmental_facts_and_statistics/world_population_clock_live/. Accessed 15 March 2021
70. GEMET (2021) Linear Economy. https://www.eionet.europa.eu/gemet/en/concept/15216. Accessed 15 March 2021
71. Brein H G (2021) How is a circular economy different from a linear economy? https://kenniskaarten.hetgroenebrein.nl/en/knowledge-map-circular-economy/difference-circular-linear-economy/. Accessed 15 March 2021
72. Barrio R (2017) Linear economy vs circular economy. https://europa.eu/capacity4dev/public-environment-climate/blog/linear-economy-vs-circular-economy. Accessed 15 March 2021
73. Ellenmacarthur Foundation (2013) Towards the circular economy. https://www.ellenmacarthurfoundation.org/assets/downloads/publications/Ellen-MacArthur-Foundation-Towards-the-Circular-Economy-vol.1.pdf. Accessed 15 March 2021
74. World Economic Forum (2014) Towards the circular economy: accelerating the scale-up across global supply chains. http://www3.weforum.org/docs/WEF_ENV_TowardsCircularEconomy_Report_2014.pdf. Accessed 15 March 2021
75. Government of the Netherlands (2021) From a linear to a circular economy. https://www.government.nl/topics/circular-economy/from-a-linear-to-a-circular-economy. Accessed 15 March 2021
76. Andersen M S (2007) An introductory note on the environmental economics of the circular economy. Sustain Sci 2:133–140. https://link.springer.com/article/10.1007%2Fs11625-006-0013-6. Accessed 15 March 2021
77. GSMArena (2021) ASUS google nexus 7. https://www.gsmarena.com/asus_google_nexus_7-4850.php. Accessed 15 March 2021
78. Ellenmacarthur Foundation (2013) Towards the circular economy. https://www.ellenmacarthurfoundation.org/assets/downloads/publications/TCE_Report-2013.pdf. Accessed 15. March 2021

79. European Commission (2011) Tackling the challenges in commodity markets and on raw materials. http://eur-lex.europa.eu/legal-content/EN/TXT/HTML/?uri=CELEX:52011DC0025&from=EN. Accessed 15 March 2021
80. Jackson T (2009) Prosperity without growth—economics for a finite planet .Earthscan Publishing, UK. https://books.google.nl/books?hl=nl&lr=&id=8C2IIPr0tMYC&oi=fnd&pg=PP2&ots=q8Tx9Ha-WS&sig=gEF1ALSx7KX9Ipj3pAAimIwqneA#v=onepage&q&f=false. Accessed 15 March.2021
81. DEFRA (2012) Resource security action plan: making the most of valuable materials. https://www.gov.uk/government/uploads/system/uploads/attachment_data/file/69511/pb13719-resource-security-action-plan.pdf. Accessed 15 March 2021
82. Williams J (2016) Plastics in a linear economy. https://makewealthhistory.org/2016/01/26/plastics-in-a-linear-economy/. Accessed 15 March 2021
83. UNEP (2011) Recycling rates of metals—a status report. https://wedocs.unep.org/handle/20.500.11822/8702. Accessed 15 March 2021
84. Kumar S, Smith SR, Fowler G, Velis C, Kumar SJ, Arya S, Rena, Kumar R, Cheeseman C (2017) Challenges and opportunities associated with waste management in India. https://www.ncbi.nlm.nih.gov/pmc/articles/PMC5383819/. Accessed 15 March 2021
85. Inderscience (2016) Consumers care about carbon footprint. https://phys.org/news/2016-02-consumers-carbon-footprint.html. Accessed 15 March 2021
86. Dockrill P (2016) Consumers have a bigger impact on the environment than anything else, study finds. https://www.sciencealert.com/consumers-have-a-bigger-impact-on-the-environment-than-anything-else-study-finds. Accessed 15 March 2021
87. GOVT.CHINADAILY.COM.CN (2020) China ssrbook 2010–2019. https://govt.chinadaily.com.cn/s/202006/29/WS5ef995fd498ed1e2f34075ef/china-statistical-yearbook-2010-2019.html. Accessed 15 March 2021
88. Sustain Your Style (2020) Fashion's environmental impact. https://www.sustainyourstyle.org/old-environmental-impacts. Accessed 31 March 2021
89. Szokan N (2016) The fashion industry tries to take responsibility for its pollution. https://www.washingtonpost.com/national/health-science/the-fashion-industry-tries-to-take-responsibility-for-its-pollution/2016/06/30/11706fa6-3e15-11e6-80bc-d06711fd2125_story.html. Accessed 31 March 2021
90. Lenzing AG (2021) Equity story—lenzing—a sustainable investment. https://www.lenzing.com/investors/lenzing-share/equity-story. Accessed 31.3.21
91. EIA (2020) Coal explained coal and the environment. https://www.eia.gov/energyexplained/index.cfm?page=coal_environment. Accessed 31 March 2021
92. Ellen Macarthur Foundation (2017) Make fashion circular. https://www.ellenmacarthurfoundation.org/programmes/systemic-initiatives/circular-fibres-initiative. Accessed 31 March 2021
93. Euromonitor International (2021) Apparel and footwear in Hong Kong, China. http://www.euromonitor.com/apparel-and-footwear. Accessed 31 March 2021
94. Morgan LR, Birtwistle G (2009) An investigation of young fashion consumers' disposal habits. Int J Consum Stud 33:190–198. https://doi.org/10.1111/j.1470-6431.2009.00756.x
95. Ellen Macarthur Foundation (2017) A new textiles economy: redesigning fashion's future. https://www.ellenmacarthurfoundation.org/assets/downloads/publications/A-New-Textiles-Economy_Full-Report_Updated_1-12-17.pdf. Accessed 31 March 2021
96. Connor M C (2017) Invisible plastic: microfibers are just the beginning of what we don't see. https://www.theguardian.com/sustainable-business/2017/jun/29/microfibers-plastic-pollution-apparel-oceans. Accessed 31 March 2021
97. Boyd R, Stern N, Ward B (2015) What will global annual emissions of greenhouse gases be in 2030, and will they be consistent with avoiding global warming of more than 2°C? http://www.lse.ac.uk/GranthamInstitute/wp-content/uploads/2015/05/Boyd_et_al_policy_paper_May_2015.pdf. Accessed 31 March 2021
98. Iea (2015) Carbon capture and storage 2015 The solution for deep emissions reductions. https://www.iea.org/publications/freepublications/publication/CarbonCaptureandStorageThesolutionfordeepemissionsreductions.pdf. Accessed 31 March 2021

99. Affine (2021) Product life cycle management in apparel industry. https://www.affineanalytics.com/blog/product-life-cycle-management-in-apparel-industry/#:~:text=According%20to%20Philip%20Kotler%2C%20'The,be%20defined%20in%203%20stages. Accessed 31 March 2021
100. Econation (2021) Circular economy. https://www.econation.co.nz/circular-economy/. Accessed 31 March 2021
101. Blackburn R (2009) Sustainable textiles: life cycle & environmental impact. Woodhead publishing Limited, UK
102. United States Environmental Protection Agency (2021) Facts and figures about materials, waste and recycling. https://www.epa.gov/facts-and-figures-about-materials-waste-and-recycling/textiles-material-specific-data#main-content. Accessed 31 March 2021
103. Smithers R (2017) UK households binned 300,000 tonnes of clothing in 2016. https://www.theguardian.com/environment/2017/jul/11/uk-households-binned-300000-tonnes-of-clothing-in-2016. Accessed 18 March 2021
104. I:CO (2021) I:CO closes loops. https://www.ico-spirit.com/en/services/. Accessed 31 March 2021
105. Bradbury M (2016) Textile recycling—closing the loop in the fashion industry. https://www.buschsystems.com/resource-center/page/textile-recycling-closing-the-loop-in-the-fashion-industry. Accessed 18 March 2021
106. Boynton J (2015) Closing the loop in fast fashion. https://www.triplepundit.com/2015/11/closing-loop-fast-fashion/. Accessed 18 March 2021
107. Econote (2016) Inditex partners with lenzing to recycle textile waste. https://advancedtextilessource.com/2016/08/05/inditex-partners-with-lenzing-to-recycle-textile-waste/. Accessed 18 March 2021
108. Braham E (2017) Closing the loop on sustainable fashion. https://www.forbes.com/sites/ashoka/2017/05/08/closing-the-loop-on-sustainable-fashion/2/#c7c8ec157e1b. Accessed 18 March 2021
109. McLean F (2017) Ashoka and C&A foundation launch social impact fund for apparel industry innovators. https://www.changemakers.com/fabricofchange/blog/ashoka-and-ca-foundation-launch-social-impact-fund. Accessed 18 March 2021
110. Campden (2021) Worn again technology: a look into the world of fabric recycling. http://www.campdenfb.com/article/worn-again-technology-look-world-fabric-recycling. Accessed 17 March 2021
111. EUR-Lex (2021) Communication from the commission to the European parliament, the council, the European economic and social committee and the committee of the regions roadmap to a resource efficient europe, COM/2011/0571 final. http://eur-lex.europa.eu/legal-content/EN/TXT/?uri=CELEX:52011DC0571. Accessed 17 March 2021
112. OCED (2021) Indicators to measure decoupling of environmental pressure from economic growth. https://www.oecd.org/env/indicators-modelling-outlooks/1933638.pdf. Accessed 17 March 2021
113. Oatshoes (2021) The world's first biodegradable shoes that bloom. http://oatshoes.com/lookbook/. Accessed 17 March 2021
114. Rhoades C (2014) There is more to closed-loop textile recycling than technological innovation. https://www.theguardian.com/sustainable-business/2014/sep/24/closed-loop-textile-recycling-technology-innovation. Accessed 18 March 2021
115. Perella M (2015) New fabrics make recycling possible, but are they suitable for high street? https://www.theguardian.com/sustainable-business/sustainable-fashion-blog/2015/jan/22/fabric-recycling-closed-loop-process-high-street-fashion. Accessed 18 March 2021
116. Shieh DJ (2013) High functional polyester polyols WO2013154874A1. https://patents.google.com/patent/WO2013154874A1/es Accessed 17 March 2021
117. Liang TM Yeater RP (1985) An advanced polyester polyol for high performance shoe sole systems. J Elastom Plastics 17(1):63–71.https://doi.org/10.1177/009524438501700106. https://www.researchgate.net/publication/249776441_An_Advanced_Polyester_Polyol_for_High_Performance_Shoe_Sole_Systems. Accessed 17 March 202

118. Coggio WD, Hevus I, Webster C, Schrock A, Thompson B, Ulrich K, Dzadek N (2015) Advancing adhesives: bio-based succinic acid polyester polyols. https://www.adhesivesmag.com/articles/93913-advancing-adhesives-bio-based-succinic-acid-polyester-polyols. Accessed 17 March 2021
119. Circle Economy (2008) Fibersort project successfully enters phase 2 of the interreg North-West Europe (Nwe) funding programme. https://www.circle-economy.com/fibersort-project-successfully-enters-phase-2-of-the-interreg-north-west-europe-nwe-funding-programme/#.WtcTOohubIV. Accessed 17 March 2021
120. Interreg (2020) Fibersort: closing the loop in the textiles industry. http://www.nweurope.eu/projects/project-search/bringing-the-fibersort-technology-to-the-market/. Accessed 17 March 2021
121. Interreg (2018) Fibersort Demo Day Recap. http://www.nweurope.eu/projects/project-search/bringing-the-fibersort-technology-to-the-market/news/demo-day-recap/. Accessed 17 March 2021
122. Rhoades C (2014) There is more to closed-loop textile recycling than technological innovation. https://www.theguardian.com/sustainable-business/2014/sep/24/closed-loop-textile-recycling-technology-innovation. Accessed 17 March 2021
123. Smits H, Cunningham G, Spathas T(2018) CIRCLETextiles closing the loop for workwear. https://www.circle-economy.com/wp-content/uploads/2017/01/ReShare-Life-Cycle-Assessment-Results.pdf. Accessed 16 April 2018
124. Smits H, Cunningham G, Spathas T(2018) CIRCLETextiles closing the loop for post consumer textile. https://www.circle-economy.com/wp-content/uploads/2017/01/Reblend-Life-Cycle-Assessment-Results.pdf. Accessed 16 April 2018
125. Deb R (2019) From manish malhotra to sabyasachi mukherjee: why buy when you can rent a LEHENGA? https://timesofindia.indiatimes.com/life-style/fashion/buzz/from-manish-malhotra-to-sabyasachi-mukherjee-why-buy-when-you-can-rent-a-lehenga/articleshow/71877383.cms. Accessed 18 March 2021
126. Hartikainen E, Laita S (2016) Sitra nominated for the world's premier circular economy award. https://www.sitra.fi/en/news/sitra-nominated-worlds-premier-circular-economy-award/. Accessed 18 March 2021
127. Tencel (2021) REFIBRA-contribution to circular economy. http://www.lenzing-fibers.com/br/tencel/refibra/. Accessed 18 March 2021

Circular Economy: An Insightful Tool for Sustainable Management of Wastewater

B. Senthil Rathi and P. Senthil Kumar

Abstract Public activities such as manufacturing, farming, and household uses have had an impact on the climate, causing serious issues such as global warming and the production of wastewater with excessive levels of chemicals. Since good-quality water is a valuable product of insufficient supply, treating wastewater to remove contaminants has become critical. Given the lack of water supplies, it is critical to comprehend and improves wastewater treatment techniques as component of water conservation. By incorporating energy efficiency and resources regeneration into the development of cleaner water, water treatment has become a component of the circular sustainability movement. In this paper, we extend the circular economy principle to the problem of wastewater management that is both sustainable and environmentally friendly. Furthermore, it has also concluded that in a circular economy method wastewater treatments system regulation is important.

Keywords Circular economy · Wastewater · Contaminants · Management · Wastewater treatment

1 Introduction

The global demand for high-quality water, either for consumption, sewage, agriculture, or manufacturing use, is already steadily increasing although there has been widespread debate in recent times regarding water treatment and reuse, which necessitates the strictest requirements [100]. Owing to population increase industrial growth, and lengthy drought the distribution of a broad variety of pollutants in groundwater and surface water is becoming a critical concern globally. Controlling the adverse effects of toxins and improving the human breathing condition are

B. S. Rathi
Department of Chemical Engineering, St. Joseph's College of Engineering, Chennai, India

P. S. Kumar (✉)
Department of Chemical Engineering, Sri Sivasubramaniya Nadar College of Engineering, Chennai, India
e-mail: senthilkumarp@ssn.edu.in

S. S. Muthu (ed.), *Circular Economy*, Environmental Footprints and Eco-design of Products and Processes, https://doi.org/10.1007/978-981-16-3698-1_7

therefore important. Heavy metals, inorganic chemicals, organic contaminants, and a variety of other diverse compounds continue to exist in wastewater [112]. Many of these toxins that are released into the atmosphere via wastewater are hazardous to humans and the atmosphere. As a result, removing toxins is becoming a necessity [21]. Faced with water scarcity, the planet is looking at all available solutions for minimizing overuse of scarce freshwater supplies. Wastewater is among the most reliable supplies of water. In order to satisfy man's weighty requirements, manufacturing, agricultural, and domestic operations expand in tandem with population growth. These operations generate vast amounts of wastewater that can be recycled and used for a variety of purposes. Traditional wastewater treatment methods have been effective in preparing effluents for disposal to some degree over the years [75].

Biochar holds promise as a treatment for wastewater, water reclamation, and gas recovery and isolation [111]. There are also several items found in nature which are of little or no value. The use of both of these resources as low-cost biosorbents for wastewater treatment can give them certain importance [35]. Filtration [39], evaporation [32, 98], ultrafiltration [70, 92], and dialysis [1, 16], coagulation [74, 105], reverse osmosis[9, 109], foam flotation [58, 90], ion exchange [23, 54], solvent extraction [40, 41], bacterial treatment [26], oxidation of chlorine [84], ozone [89], hydrogen peroxide [50, 113], and chlorine dioxide [2, 42], aerobic and anaerobic treatment [25, 43], photochemical reactions [24, 91], electro dialysis [36, 65], activated sludge [30], microbial reduction [110], electrochemical treatment [60] and ultrasonic treatment [63, 69].

For the most cost-effective maintenance of drainage systems, there is a growing agreement that a strategic management scheme, wherein steps are prepared and done until the program fails functionally, should be implemented. In particular, conventional wastewater treatment schemes are proactive based on a "repair it whenever it breaks" technique [7]. Growing consumer demands on privately owned water utilities, tougher environmental laws, and imposition of suburban neighborhoods on water treatment plants have ended in a spike in the amount of consumer pollution grievances in recent years [55].

A common philosophy advocated by several state governments and many enterprises across the globe is the circular economy. Even so, the circular economy concept's science and analysis material is shallow and disorganized. Circular economy appears to be a jumble of nebulous and disparate theories from various disciplines, as well as semi-scientific principles [47, 49]. Although the words circular economy and sustainable development are increasing popularity among academics, business, and politicians, the parallels and distinctions between the two definitions are still unclear. The connection between the principles is not explicitly stated in the literature, which blurs their philosophical shapes and limits the effectiveness of using the methods in study and practice [28]. In light of legislative criteria and circular economy concepts, using wastewater slurry as a part of renewable and energy reuse is a viable option for its management. Scientists are considering the removal of essential components from sediment, like carbon and nutrients, now that it is being seen as an asset rather than a loss. The electricity that can be extracted from drainage sludge may be a long-term option for meeting current and potential energy needs [31].

This chapter discusses wastewater and its origins, wastewater disposal, wastewater management, circular economy, and the relationship between circular economy and wastewater management.

2 Wastewater Pollutants

High-quality water is necessary for human survival, and water of good standard is required for irrigation, industry, residential, and business reasons. Many of these practices contribute to the pollution of the water. Any day, billions of gallons of sewage from all of these outlets were dumped into natural waters. The demand for freshwater is although all water supplies are gradually being inadequate for use as a result of insufficient waste management. The challenge of supplying adequate treatment facilities for all sources of pollution is daunting and costly, and there is an urgent need for low-cost, low-maintenance, and energy-efficient advanced technology [88]. Water contamination has risen dramatically as a result of ever-increasing industrial development and accelerated urban development. Environmental conservation authorities have implemented more restrictive legislative restrictions and, in collaboration with certain non-governmental groups, have begun to increase their vigilance in order to preserve the ecosystem. This has increased the cost of water treatment, and meeting the discharged efficiency requirement has been a massive challenge for the factories. As a result, it was thought that the potential for water recycling for different uses must be explored [94].

Pollution sources are classified into two categories: point sources and non-point sources. Localized recognizable origins of pollution, such as electricity stations, refineries, mining, mills, wastewater treatment plants, and so on, are known as point sources. Non-point sources, including a river, are those that are dispersed over a large geographic region. Mobile sources such as vehicles, taxis, and trains are examples of non-point sources. Even though each one of these is a single point origin, these are all shifting and therefore have a significant regional influence. Urban drainage is a good example of a non-point source of waste since the contaminant load can be the result of thousands of small point sources. Anthropogenic sources of organic chemical pollutants, marine debris, and plastic in the environment, metals and metalloids, nutrients, radionuclides, bacterial contamination, and other water pathogens were various sources of contaminants in wastewater [95].

Increased heavy metal discharge into the atmosphere as a result of urbanization and industrialization has been a major issue around the globe. Apart from chemical compounds, who dissolve into harmless products in the number of cases toxic heavy metals do not dissolve into harmless products. Because of its toxicity to several forms of life, the existence of heavy metal ions is a significant topic of problems. Metal plating, mine activities, dairies, chloralkali, radiator processing, smelters, alloy plants, and storage battery industries all have heavy metal contaminants in their aqueous wastes [38]. Water supplies have been shown to be contaminated by a few hundred chemical toxins. Organic pollutant toxicity is very harmful owing to their many adverse effects and cancer causing existence. As a result, maintaining and improving water quality is critical in everyday life, and the value of doing so is growing. Water contamination is a major concern for researchers, academics, and government officials around the world. Several countries around the world have polluted surface water and ground waters that are unsafe to drink [4].

A number of chemicals have been used in everyday activities of our current life. Synthesized chemicals are found in foods, additives, perfumes, cosmetics, adhesives, cleaning supplies, plastics, paintings, and other manufacturing, and consumer goods. Many of these substances shield people and animals from deadly diseases, boost food crop productivity, and enhance our quality of life in general. While beneficial to humans, a few of these contaminants cause water pollution and change environmental protection, altering the health of plants and animals [62]. Some of the other pollutants are biological agents, heat, dissolved and non-dissolved chemicals, heavy metals, dyes, phenols, radioactive substances, detergents, polychlorinated biphenyls, pesticides, and other organic chemicals [35].

3 Treatment Process of Wastewater

Freshwater has also been found to contain over 700 chemical compounds, which come from a variety of urban, manufacturing, and farming processes, and also natural decay of animal or vegetable waste [79]. Quite small suspended solids, dissolved solids, natural, and inorganic substances, metals, as well as other contaminants are commonly found in wastewater provided by various manufacturers. Due to the limitations of the samples and the existence of charge density, bringing them together to form a heavier mass for settlement and filtration is difficult. As a result, factories face a significant obstacle in removing colloidal particles from wastewater. To isolate particles from wastewater, a variety of conventional and modern technology has been used, including membrane filtration, ion exchange, flotation, precipitation, solvent extraction, flocculation, adsorption, coagulation, biochemical, and electrolytic methods [56].

Adsorption is the mechanism by which a liquid solute collects on a solid's surface and creates an atomic or molecular film. There are several different types of adsorbent materials that can be used in their original or changed states to remove toxic metal ions from drainage effluents. Activated carbons, clay minerals, biomaterials, zeolites, and municipal solid waste are the most often used. For the extraction of natural and synthetic micropollutants from the aqueous process, adsorption technology is now widely used [12, 52]. The most common adsorbent is activated carbon. It's a translucent, undifferentiated solid made up of small crystals with a graphite structure that's usually sold as tiny pieces or powder form. It is capable of removing a wide range of hazardous metals [53]. When it came to agricultural and domestic waste, banana and orange peels had a stronger preference for removing dyes, heavy metals, and other contaminants like phosphorous and nitrogen compounds. Furnace sludge, carbon-based adsorbents from the fertilizers, coal ash, and red mud are the best among industrial materials in this regard. Still, fly ash, and other agricultural waste products, may be used with several other methods to substitute conventional activated charcoal in some cases [20, 46, 93]. It is worth noting that almost all agriculture-based waste is no longer used in its natural form but is now altered in a number of ways to increase porosity and adsorption surface area [96]. Nano structuring, activation, combustion, and gluing are some of the most popular agricultural residues sorbents alteration techniques [44, 68, 106]. Owing to the easily manipulated binding sites, precise surface region, pore depth, pore size distribution, fast isolation, and recyclability, magnetic bio sorbents have drawn growing research interest. They are ideal for the abatement of heavy metal and organic pollutants [13, 29, 51]. The characteristics of magnetic adsorbents are influenced by the biomass feedstock, metal nanoparticle characteristics, modify processes, and processing methods, all of which affect the efficiency of organic and inorganic pollutants removal speed [37, 81]. Among all current technology for water treatment, adsorption is a simple, long-lasting, cost-effective, and environmentally friendly process. However, more research funding, optimization, and functional deployment of the integrated method for a variety of applications are needed [86, 101]. In contrast to many other adsorbent materials such as commercial activated carbon and thermally and chemically treated biochar, magnetic adsorbents are successful adsorbent materials in terms of their reduction ability, fast and simple magnetic isolation, and numerous recycling to reduce the cost of abatement of organic and inorganic pollutants from wastewater [80, 83, 86].

Ion exchange is a flexible isolation method with a wide range of uses in water quality management. For several years, the ion exchange mechanism has also been economically feasible. The wide variety of resins available today is a driving force behind curiosity in ion exchange technologies. Ion exchange will have an efficient and cost-effective approach to emissions control needs with sufficient resin choice. As water with soluble electrolytes comes into contact with such solids, it improves. Mostly on surface or even within the solid ion exchange resin, any of the particles found in the sample will exchange for other ions [11]. Ion exchange systems usually produce ion exchange resins that work in a systematic way. Water passes through the base until it is saturated, after which point the water that emerges from the resin has a higher intensity of the ions that must be separated. After backwashing the resin to dissolve the dissolved solids, the resin is recovered by rinsing the expelled ions from the resin with a solution of substitute resin. As a result, the use of ion exchange in water treatment is limited due to the creation of backwash [45]. Ion exchange as a water remediation strategy has the benefit of being very economically efficient. Resin regeneration requires small amount of energy and is very cost-effective. If resins are well treated, they can survive for several years before needing to be substituted removal efficiencies of heavy metals in the concentrations spectrum present in wastewater effluents and sludge is nearly 99.9%. Selective heavy metal treatment from polluted wastewater/sludge incorporates the advantages of being cost-effective and allowing the processed wastewater effluents and sewage sludge to be reused or recycled [3].

Membrane technology's use in water treatment has grown in recent decades as a result of more rigorous regulations and advancements in membrane process [27]. The membrane bioreactor method, in specific, has been commonly used as an alternative to traditional activated sludge process for the isolation and preservation of biological solid particles [14]. Several substances in the wastewater will accumulate on the membrane during membrane distillation process, resulting in deposits. In the vast bulk, of instances this phenomena has a substantial impact on process efficiency, ruling out the use of membrane distillation for specific water treatment. If wastewater is pre-treated properly, this problem can be avoided. The ability to remove foulants from the feed is the primary benefit of the membrane distillation process for wastewater treatment [33]. For the rehabilitation of a wide variety of complicated industrial wastes, reverse osmosis was shown to be a successful and cost-effective process stage. Because of the popularity of reverse osmosis in large-scale desalination and urban wastewater treatment, many companies now see this technique as a way to reduce emissions and save money by reusing it [22, 102]. Pilot-scale implementation of membrane processes to specialized textile water treatment, downstream of a biologically sludge operation, with the goal of reusing water in textile technology processes. Sand filtration was used as a pre-treatment, followed by a microfiltration or ultrafiltration membrane procedure, and a final separation treatment using a Nanofiltration or reverse osmosis membrane [64].

Electrochemical technology has excellent means for dealing with ecological issues. The key reagent being used is electrons, which really is a cleaner reagent, so no additional reagent is needed [10, 85]. As an appropriate alternative approach for treating wastewater, the electrochemical method has been shown to be very efficient and cost-effective [59, 78]. Pollutants are eliminated in the electrochemical process through either direct or indirect oxidation. The contaminants first were deposited on the anode and then eliminated by the anodic electron transfer process throughout the direct anodic oxidation method. Strong oxidizing agents such as hypochlorite/chlorine, ozone, and hydrogen peroxide are electrochemically produced in the indirect oxidation method. The contaminants are then killed in the solution phase by the produced oxidant's oxidation reaction. Many of the oxidants are produced on-site and used right away [5, 85]. Over more than half a century, the electro dialysis method has been well established in water purification as a proprietary operation. Ion exchange membranes compete with other isolation methods such as reverse osmosis, ultrafiltration, Nanofiltration, and traditional ion exchange in most cases. Electro dialysis methods, according to a few research, have more benefits than other separation procedures. There is no osmotic strain, the substance is of better quality, it is ecologically sound, it requires no extra additives, and the ion exchange membranes have quite a long lifespan. However, electro dialysis has a number of drawbacks [77, 87]. Membrane fouling is a significant drawback of the method because it lowers the limiting current, flux, membrane resistance, ions movement yield and causes severe depletion issues. Fouling rises as fluid flow decreases, current rises, and colloid intensity rises [76]. Cleanup of hazardous wastewaters has become a top priority due to public pressure for pollutant-free dumping waste to receiving water bodies. This is, however, a complex and difficult mission. It is still difficult to come up with a single formula for getting rid of all toxins from wastewater. Even so, just a handful of the many and varied wastewater treatment commonly listed are widely used by the manufacturing industry for financial and strategic reasons. Despite this, adsorption into activated carbon is often quoted as the preferred method for removing a variety of toxins because it has the best outcomes in terms of efficiency and technological feasibility on an industrial scale [19].

4 Management of Wastewater

The primary purpose of wastewater management is environmental conservation that is proportional to human health and social considerations. It is proposed whether main, intermediate, or tertiary treatment would be taken in before final discharge based on the quality of the wastewater. Knowing the essence of wastewater is essential for designing an effective wastewater treatment method, adopting an appropriate protocol, determining suitable residue requirements, determining the degree of assessment needed to verify the procedure, and deciding on the residues to be evaluated based on toxicity, among other things. As a result, it's vital to maintain the processed wastewater's protection, efficiency, and consistency [107]. Because of the high levels of pollution and the extent of scarcity, a need for long-term water supply management is becoming increasingly important. Even so, the world is repeatedly subjected to extremely stressful conditions caused by inadequate or non-existent wastewater and wastewater treatment facilities, restricting access to sanitation, and causing health problems as a result. To address this issue, decentralization, in conjunction with municipal government, is being progressively recognized as a possibly viable means of decreasing the number of people without exposure to safe drinking water or adequate sanitation, and also the quality of water treatment and treated wastewater reuse and recycling. To achieve environmental efficiency, all pollution dilution processes should be minimized, and treated wastewater reuse and by-products recycling should be maximized. Treatment methods can be effective and dependable, with low design, management, and repair costs that promote self-sufficiency and community acceptance [57].

Wastewater management is an essential step in reducing pollution and improving the efficiency of the water supply. The treating wastewater scheme is unpredictable and variable due to the complexities of environmental environments, influent shock, and wastewater treatment technologies. These risks lead to changes in effluent quality of water and operating costs, as well as the possibility of receiving waters becoming contaminated. Artificial intelligence has evolved into a potent method for reducing the uncertainties and problems associated with wastewater treatment [114]. To ensure safe and intelligent control of water supply systems, innovative techniques and technology are needed to address the worsening water climate. Wireless sensor are a promising tool for tracking and managing quality of the water. Wireless sensor make it easier to upgrade existing centralized processes and manual processes, resulting in integrated smart water quality control systems that can respond to cities' complex and varied water delivery networks. However, there is a commercial need for a low-cost wireless sensor network model that enables the introduction of this latest generation of devices at a reasonable cost [66]. Wastewater management can be comprehensive, focusing on conscientious and collaborative assessment of waste collection, storage, disposal, and rehabilitation. Pollution should be contained, and the wastewater management area must be held to the smallest feasible scale practicable, with wastes filtered as few as probable. Wastewater must be treated as near to its points as possible, and where it could be reused; waste exports must be reduced to increase productivity and limit contamination distribution [8].

5 Circular Economy

The circular economy is by far the most recent effort to comprehend a sustainable technique of integrating economic growth and ecological well-being [49]. Regardless of the fact that the circular economy is becoming more common as a business concept, there is very no systematic scholarly discussion about it in the sustainability into business literature. Nonetheless, considering the multitude of scholarly discussions on ecological and environmentally beneficial business wherein financial growth is often seen to be prioritized above financial, social, and moral and ethical principles, the circular economy is an important phenomenon to investigate [47]. The circular economy is a major recent way of thinking in environmental sustainability, having been embraced as the key paradigm for ecological inflation and technological growth by the world's largest country, China, over the next ten years. A circular economy acknowledges that people, their actions, and their world all are loci on the very same ring. A circular economy entails whole distribution networks, and liability is distributed across these networks, with neither the manufacturer nor the customer being relatively predictable [73]. A circular economy has the ability to contribute to long-term sustainability by detaching economic development from resources depletion and environmental deterioration. Despite its importance in today's policy and economic debates, the idea of a circular economy is still subject to interpretation [71]. Despite the trend toward growth in the economy, a transition to a circular economy necessitates eco-innovations to close the circle of a product life cycle, divert essential products from pollution, and address the needs of ecological sustainability. "The development, use, or exploitation of a product, commodity, manufacturing process, organization culture, or administration or business process that is new to the company or consumer and that outcomes, during its lifespan, in a reduced environmental risk, contamination, and unfavorable resources utilize (including energy use) relative to pertinent" is how eco-innovation is described. The study of the circular economy hypothesis led to some other significant discovery: the creation of an information mapping that demonstrates how the circular economy is the result of three phases of societal, industrial, and economic developments that are all linked to how society develops new. However, it explicitly reflects the paradigm shift's most mature and recent representation. Issues over conservation purposes, ethics, and safety, as well as greenhouse-gas decreases, are causing us to view commodities as properties that should be stored rather than eaten indefinitely. This ensures that the 3R concepts should be implemented during the whole period of resource creation, use, and returns, whereas sustainable engineering policies serve as catalysts and guidance for creating products and services that can be reintroduced into the economy as biological or technological tools in the long run [82].

6 WasteWater Management and Its Circular Economy

Insufficient clean drinking water and sanitation are among the most prevalent issues afflicting people around the world. Water is becoming a limited commodity in terms of quantities and efficiency as the world economy and population expand. The water problem has been identified by the World Economic Forum as the global danger with the greatest potential for devastation. Water restoration at wastewater treatment facilities is an integral aspect of long-term water supply conservation in this sense. Disposal of wastewater may be used for a variety of advantageous uses including farm drainage, manufacturing operations, water reuse, and, after efficient technology, also potable water supply. Besides that, the ever-increasing need for a much more productive society is driving new advancements in sewage treatment, with the aim of recovering all usable resources in wastewater (e.g., renewable water, manure, energy, bio-plastics, and other materials), also in the context of on-site recycling in line with the approved "circular economy" model [97]. The circular economy can be seen in the water treatment industry as well. Consideration of wastewater restoration and recycling, for example, seems to be an ideal choice for increasing water supplies while reducing the carbon footprint [34].

Water storage, reuse, and processing of polluted water from non-traditional supplies to "make" freshwater would be needed to meet humankind's ever-increasing need for sources, such as water, electricity, and food. Recycling and reuse are key components of a circular economy plan, and they provide a way to increase water supply by properly treating wastewater. Recycled water faces various obstacles, ranging from public sentiment to pricing and regulatory issues, all of which could be more easily tackled if approached from a circular economy approach [108]. In certain situations, sustainable water supply management is insufficient to meet circular economy goals because manufacturing, utility, and domestic operations also produce vast volumes of wastewater which must be discarded of. As a result, wastewater treatment, which is an essential aspect of waste control, is given particular consideration. It includes the six activities mentioned below that are helpful in implementing circular economy principles in the water and wastewater industry: (i) Reduction is the process of preventing wastewater production during the first place by reducing water use and reducing emissions at the origin, (ii) Restoration is the use of effective methods to remove toxins from water and sanitation, (iii) Reuse of treated wastewater as an additional source of non-potable water system, (iv) Recycling is the process of recovering potable water from wastewater, (v) Recovery is services such as nutrients and electricity may be recovered from water-based waste, and (vi) Rethink when to use energy to build a waste-free, competitive economy [104].

Nutrient restoration can be thought of as a chain or process that comprises the following stages: (i) agriculture; (ii) handling; (iii) usage; (iv) waste and wastewater collection; and (v) waste and wastewater disposal. The cycle is completed as the minerals reclaimed from irrigation are returned to agriculture as mineral and organic compounds. As a result, urban wastewater serves as an important secondary source of nutrients. As an alternative, Clinoptilolite is used for adsorption (zeolite) Potassium, phosphorus, and ammonium Orthophosphate removal efficiencies ranged from 64 to 80%, ammonium removal efficiencies from 40% to 89 percent, and potassium removal efficiencies from 37 to 78%. Recovery time is high to moderate. Operation is easy [48]. The food processing industry generates large quantities of nutrient-rich pollutants that could be used as a tool to recycle high-value materials, allowing for the development of much more ecologically responsible technologies. Polyphenols are divided among water and oil processes during the production of olive oil, with the key component being found in wastewater due to its higher polarization and solubility in water. Polyphenol levels in olive oil effluents ranged from 5 to 25 g/L on average. As natural antioxidants of great importance to the dairy, medicinal, and cosmetic industries, their recovery is promising [67]. Because of its increased nutrient load and poor dewaterability, waste sludge treatment and disposal is a major environmental issue, with high operating costs. Furthermore, it is becoming a global issue as the number of wastewater treatment plants increases across the world, resulting in an increase in sludge demand. In 2015, for example, over 10 million tons of dry solids of sludge were generated in the EU. This must be, combined with strengthening of ecological quality requirements, and necessitates proper waste management. As a result, effective management is critical to lowering maintenance costs and making wastewater disposal more environmentally sustainable. Most countries are concentrating on sludge recoveries, such as using it to generate electricity by incineration or anaerobic digestion [18].

Enabled sludge contains a variety of heavy metals, including Cu, Ni, Zn, Cd, Pb, Cr, and Hg, which limits it is used for irrigation purposes due to the risk of soil and groundwater pollution, which can be harmful to human and animal health. It is worth noting that the concentrations of these components vary considerably depending on the source of sewage sludge, with biosolids having the maximum concentrations. As a result, sludge must be processed to recover heavy metals before it can be valorized [72].

Because of their simplicity and ease of use, adsorption processes have received a lot of attention. Furthermore, through use of waste adsorbents is receiving more consideration in order to improve low-cost water management solutions. Precursors may include treated effluent sewage sludge, dewatered sewage sludge, and sewage sludge from municipal/urban or WWTPs. For adsorbent preparation, various methods have been proposed, the most commonly used being (i) carbonization, (ii) physical activation, (iii) chemical activation, and (iv) a mixture of physical and chemical activation [103].

Since the quality of sewage sludge oxides (Al_2O_3, CaO, SiO_2, and Fe_2O_3) is close to that of Cement concrete or clay, it has been suggested that it be used to make building and construction products such as environmentally sustainable, tiles, ceramic, cementitious materials, or durable aggregates. However, sewage sludge contains a high volume of organic matter, which can damage the cementitious properties, resulting in poor bonding strength. As a result, sewage sludge must be pre-treated before being used in the production of cement or concrete [15]. Because of their usage as fluid fertilizer application protein foams, glues, and livestock feed, proteins may be called high-added-value items. Despite the high percentage of protein (up to 61%) in activated sludge and the fact that crude protein accounts for about 50 percent of the dry mass of bacterium, protein restoration is a hot topic. In this regard, various protein recovery methodologies have been suggested, with the first stage often being the solubilization of intracellular content in the sludge [17].

Depending on the scale and the procedures it does, a wastewater treatment facility absorbs 0.45–1.25 kWh/m^3. This electricity is normally supplied to the plant from outside sources, such as traditional power plants. Municipal wastewater, on the other hand, has a gross energy content of up to 9.7 kWh/m^3. If the total energy consumption of wastewater treatment plants is 0.85 kWh/m^3, the water will produce up to 12 times the amount of energy used for treatment. This power in sewage can be classified as chemical (from the organic matters in the water), thermal, or energy stored, and it can be extracted from the water using various technologies [61].

Case studies; In preparation to become personality in terms of water supplies, Singapore's Public Utilities Board manufactures high-grade recycled water known as NEWater, which accounts for 30% of the city-water state's needs. NEWater is created using a sophisticated cleaning procedure that includes conventional wastewater treatment in a water reclamation facility, microfiltration/ultrafiltration, and reverse osmosis. The water is now of drinking water quality; UV disinfection will guarantee that any remaining microbes are rendered inactive. It has been used in industry for reservoir augmentation at a $1.22 per gallon water tariff.

The Virginia Irrigated agriculture Recycling Scheme, founded in 1999, is Australia's very first biggest water treatment scheme, supplying around 20 gigaliters of highly treated greywater each year to 400 linkages in the Virginia and Angle Vale regions north of Adelaide via a vast system of valves. This water is provided by a dissolved air floatation/filtration plant that is supplied by processed sewage from either the Bolivar Sewage Treatment Plant, bringing the Bolivar plant's reuse rate from around 29% to 35%. It was used in irrigation at a $0.095 water tariff [6].

7 Conclusion

Natural and artificial pollutants are generated when organic and inorganic compounds are introduced into the atmosphere as a result of residential, agricultural, and industrial water operations. The standard primary and secondary treatment systems for these wastewaters have been implemented in an increasing number of locations in order to remove readily settled substances and oxidize the organic matter present in wastewater. Household agricultural and manufacturing waters are among the pollutant sources. Because of structural, societal, and economic hurdles, wastewater reuse is not well developed. The circular economy's implementation could lead to more accurate and safe water usage [99]. For a long-term water reuse implementation, a comprehensive approach that considers all reuse variables such as political, decisional, social, fiscal, technical, and environmental factors is required. In reality, after defining the characteristics of a particular model, present study tested its relevance by applying it to the actual life. Abnormalities can occur during the application, resulting in a paradigm shift and, as a result, a "science revolution." As a result, there is likely a need to implement a "technical evolution" that will lead to more efficient wastewater management and improved resource recovery.

References

1. Agar JW (2008) Reusing dialysis wastewater: the elephant in the room. Am J Kidney Dis 52(1):10–12. https://doi.org/10.1053/j.ajkd.2008.04.005
2. Aieta EM, Berg JD (1986) A review of chlorine dioxide in drinking water treatment. J Am Water Works Ass 78(6):62–72. https://doi.org/10.1002/j.1551-8833.1986.tb05766.x
3. Al-Enezi G, Hamoda MF, Fawzi N (2004) Ion exchange extraction of heavy metals from wastewater sludges. J Environ Sci Health, Part A 39(2):455–464. https://doi.org/10.1081/ESE-120027536
4. Ali I, Asim M, Khan TA (2012) Low cost adsorbents for the removal of organic pollutants from wastewater. J Environ Manage 113:170–183. https://doi.org/10.1016/j.jenvman.2012.08.028
5. Alsheyab M, Jiang JQ, Stanford C (2009) On-line production of ferrate with an electrochemical method and its potential application for wastewater treatment–A review. J Environ Manage 90(3):1350–1356. https://doi.org/10.1016/j.jenvman.2008.10.001
6. Asolekar S, Biniwale RB, Chandran K, Chaudhuri RR, Heemskerk F, Jain AK et al. (2016) Circular economy pathways for municipal wastewater management in India: a practitioner's guide. Circular economy pathways for municipal wastewater management in India: a practitioner's guide
7. Baik HS, Jeong HS, Abraham DM (2006) Estimating transition probabilities in Markov chain-based deterioration models for management of wastewater systems. J Water Resour Plan Manag 132(1):15–24
8. Bakir HA (2001) Sustainable wastewater management for small communities in the Middle East and North Africa. J Environ Manage 61(4):319–328. https://doi.org/10.1006/jema.2000.0414
9. Bartels CR, Wilf M, Andes K, Iong J (2005) Design considerations for wastewater treatment by reverse osmosis. Water Sci Technol 51(6–7):473–482. https://doi.org/10.2166/wst.2005.0670
10. Basha CA, Ramanathan K, Rajkumar R, Mahalakshmi M, Kumar PS (2008) Management of chromium plating rinsewater using electrochemical ion exchange. Ind Eng Chem Res 47(7):2279–2286. https://doi.org/10.1021/ie070163x
11. Bochenek R, Sitarz R, Antos D (2011) Design of continuous ion exchange process for the wastewater treatment. Chem Eng Sci 66(23):6209–6219. https://doi.org/10.1016/j.ces.2011.08.046
12. Burakov AE, Galunin EV, Burakova IV, Kucherova AE, Agarwal S, Tkachev AG, Gupta VK (2018) Adsorption of heavy metals on conventional and nanostructured materials for

wastewater treatment purposes: a review. Ecotoxicol Environ Saf 148:702–712. https://doi.org/10.1016/j.ecoenv.2017.11.034

13. Carolin CF, Kumar PS, Joshiba GJ, Madhesh P, Ramamurthy R (2021) Sustainable strategy for the enhancement of hazardous aromatic amine degradation using lipopeptide biosurfactant isolated from Brevibacterium casei. J Hazardous Mater 408:124943. https://doi.org/10.1016/j.jhazmat.2020.124943
14. Chang IS, Kim SN (2005) Wastewater treatment using membrane filtration—effect of biosolids concentration on cake resistance. Process Biochem 40(3–4):1307–1314. https://doi.org/10.1016/j.procbio.2004.06.019
15. Chang Z, Long G, Zhou JL, Ma C (2020) Valorization of sewage sludge in the fabrication of construction and building materials: a review. Resources, Conservation Recycling 154:104606. https://doi.org/10.1016/j.resconrec.2019.104606
16. Chen C, Dong T, Han M, Yao J, Han L (2020) Ammonium recovery from wastewater by donnan dialysis: a feasibility study. J Cleaner Prod 265:121838. https://doi.org/10.1016/j.jclepro.2020.121838
17. Chen X, Li C, Ji X, Zhong Z, Li P (2008) Recovery of protein from discharged wastewater during the production of chitin. Biores Technol 99(3):570–574. https://doi.org/10.1016/j.biortech.2006.12.029
18. Cieślik BM, Namieśnik J, Konieczka P (2015) Review of sewage sludge management: standards, regulations and analytical methods. J Clean Prod 90:1–15. https://doi.org/10.1016/j.jclepro.2014.11.031
19. Crini G, Lichtfouse E (2019) Advantages and disadvantages of techniques used for wastewater treatment. Environ Chem Lett 17(1):145–155. https://doi.org/10.1007/s10311-018-0785-9
20. De Gisi S, Lofrano G, Grassi M, Notarnicola M (2016) Characteristics and adsorption capacities of low-cost sorbents for wastewater treatment: a review. Sustain Mater Technol 9:10–40. https://doi.org/10.1016/j.susmat.2016.06.002
21. Deblonde T, Cossu-Leguille C, Hartemann P (2011) Emerging pollutants in wastewater: a review of the literature. Int J Hyg Environ Health 214(6):442–448. https://doi.org/10.1016/j.ijheh.2011.08.002
22. Dialynas E, Mantzavinos D, Diamadopoulos E (2008) Advanced treatment of the reverse osmosis concentrate produced during reclamation of municipal wastewater. Water Res 42(18):4603–4608. https://doi.org/10.1016/j.watres.2008.08.008
23. Eom TH, Lee CH, Kim JH, Lee CH (2005) Development of an ion exchange system for plating wastewater treatment. Desalination 180(1–3):163–172. https://doi.org/10.1016/j.desal.2004.11.088
24. Espíndola JC, Cristóvão RO, Araújo SR, Neuparth T, Santos MM, Montes R et al (2019) An innovative photoreactor, FluHelik, to promote UVC/H2O2 photochemical reactions: Tertiary treatment of an urban wastewater. Sci Total Environ 667:197–207. https://doi.org/10.1016/j.scitotenv.2019.02.335
25. Fricke K, Santen H, Wallmann R (2005) Comparison of selected aerobic and anaerobic procedures for MSW treatment. Waste Manage 25(8):799–810. https://doi.org/10.1016/j.wasman.2004.12.018
26. Gallert C, Winter J (2005) Bacterial metabolism in wastewater treatment systems. Environ Biotechnol Concepts Appl. https://doi.org/10.1002/3527604286
27. Gebreyohannes AY, Mazzei R, Giorno L (2016) Trends and current practices of olive mill wastewater treatment: application of integrated membrane process and its future perspective. Sep Purif Technol 162:45–60. https://doi.org/10.1016/j.seppur.2016.02.001
28. Geissdoerfer M, Savaget P, Bocken NM, Hultink EJ (2017) The Circular Economy–a new sustainability paradigm? J Clean Prod 143:757–768. https://doi.org/10.1016/j.jclepro.2016.12.048
29. Gerard N, Krishnan RS, Ponnusamy SK, Cabana H, Vaidyanathan VK (2016) Adsorptive potential of dispersible chitosan coated iron-oxide nanocomposites toward the elimination of arsenic from aqueous solution. Process Saf Environ Prot 104:185–195. https://doi.org/10.1016/j.psep.2016.09.006

30. Gernaey KV, van Loosdrecht MC, Henze M, Lind M, Jørgensen SB (2004) Activated sludge wastewater treatment plant modelling and simulation: state of the art. Environ Model Softw 19(9):763–783. https://doi.org/10.1016/j.envsoft.2003.03.005
31. Gherghel A, Teodosiu C, De Gisi S (2019) A review on wastewater sludge valorisation and its challenges in the context of circular economy. J Clean Prod 228:244–263. https://doi.org/10.1016/j.jclepro.2019.04.240
32. Green M, Shaul N, Beliavski M, Sabbah I, Ghattas B, Tarre S (2006) Minimizing land requirement and evaporation in small wastewater treatment systems. Ecol Eng 26(3):266–271. https://doi.org/10.1016/j.ecoleng.2005.10.007
33. Gryta M, Tomaszewska M, Karakulski K (2006) Wastewater treatment by membrane distillation. Desalination 198(1–3):67–73. https://doi.org/10.1016/j.desal.2006.09.010
34. Guerra-Rodríguez S, Oulego P, Rodríguez E, Singh DN, Rodriguez-Chueca J (2020) Towards the implementation of circular economy in the wastewater sector: challenges and opportunities. Water 12(5):1431. https://doi.org/10.3390/w12051431
35. Gupta VK, Carrott PJM, Ribeiro Carrott MML, Suhas (2009) Low-cost adsorbents: growing approach to wastewater treatment—a review. Critical Rev Environ Sci Technol 39(10):783–842. https://doi.org/10.1080/10643380801977610
36. Gurreri L, Tamburini A, Cipollina A, Micale G (2020) Electrodialysis applications in wastewater treatment for environmental protection and resources recovery: a systematic review on progress and perspectives. Membranes 10(7):146. https://doi.org/10.3390/membranes10070146
37. Hassan M, Naidu R, Du J, Liu Y, Qi F (2020) Critical review of magnetic biosorbents: Their preparation, application, and regeneration for wastewater treatment. Sci Total Environ 702:134893. https://doi.org/10.1016/j.scitotenv.2019.134893
38. Hegazi HA (2013) Removal of heavy metals from wastewater using agricultural and industrial wastes as adsorbents. HBRC J 9(3):276–282. https://doi.org/10.1016/j.hbrcj.2013.08.004
39. Hollender J, Zimmermann SG, Koepke S, Krauss M, McArdell CS, Ort C et al (2009) Elimination of organic micropollutants in a municipal wastewater treatment plant upgraded with a full-scale post-ozonation followed by sand filtration. Environ Sci Technol 43(20):7862–7869. https://doi.org/10.1021/es9014629
40. Hu G, Li J, Hou H (2015) A combination of solvent extraction and freeze thaw for oil recovery from petroleum refinery wastewater treatment pond sludge. J Hazardous Mater 283:832–840. https://doi.org/10.1016/j.jhazmat.2014.10.028
41. Hu H, Yang M, Dang J (2005) Treatment of strong acid dye wastewater by solvent extraction. Separation Purif Technol 42(2):129–136. https://doi.org/10.1016/j.seppur.2004.07.002
42. Huber MM, Korhonen S, Ternes TA, Von Gunten U (2005) Oxidation of pharmaceuticals during water treatment with chlorine dioxide. Water Res 39(15):3607–3617. https://doi.org/10.1016/j.watres.2005.05.040
43. Joss A, Andersen H, Ternes T, Richle PR, Siegrist H (2004) Removal of estrogens in municipal wastewater treatment under aerobic and anaerobic conditions: consequences for plant optimization. Environ Sci Technol 38(11):3047–3055. https://doi.org/10.1021/es0351488
44. Jothirani R, Kumar PS, Saravanan A, Narayan AS, Dutta A (2016) Ultrasonic modified corn pith for the sequestration of dye from aqueous solution. J Indus Eng Chem 39:162–175. https://doi.org/10.1016/j.jiec.2016.05.024
45. Kansara N, Bhati L, Narang M, Vaishnavi R (2016) Wastewater treatment by ion exchange method: a review of past and recent researches. Environ Sci Indian J 12(4):143–150
46. Kaveeshwar AR, Kumar PS, Revellame ED, Gang DD, Zappi ME, Subramaniam R (2018) Adsorption properties and mechanism of barium (II) and strontium (II) removal from fracking wastewater using pecan shell based activated carbon. J Cleaner Prod 193:1–13. https://doi.org/10.1016/j.jclepro.2018.05.041
47. Kirchherr J, Reike D, Hekkert M (2017) Conceptualizing the circular economy: an analysis of 114 definitions. Resources, Conservation Recycling 127:221–232. https://doi.org/10.1016/j.resconrec.2017.09.005

48. Kocatürk-Schumacher, N. P., Bruun, S., Zwart, K., & Jensen, L. S. (2017). Nutrient recovery from the liquid fraction of digestate by clinoptilolite. CLEAN–Soil, Air, Water, 45(6), 1500153. https://doi.org/10.1002/clen.201500153
49. Korhonen J, Honkasalo A, Seppälä J (2018) Circular economy: the concept and its limitations. Ecol Econ 143:37–46. https://doi.org/10.1016/j.ecolecon.2017.06.041
50. Ksibi M (2006) Chemical oxidation with hydrogen peroxide for domestic wastewater treatment. Chem Eng J 119(2–3):161–165. https://doi.org/10.1016/j.cej.2006.03.022
51. Kumar PS, Ngueagni PT (2021) A review on new aspects of lipopeptide biosurfactant: Types, production, properties and its application in the bioremediation process. J Hazardous Mater 407:124827. https://doi.org/10.1016/j.jhazmat.2020.124827
52. Kumar PS, Pavithra J, Suriya S, Ramesh M, Kumar KA (2015) Sargassum wightii, a marine alga is the source for the production of algal oil, bio-oil, and application in the dye wastewater treatment. Desalination Water Treatment 55(5):1342–1358. https://doi.org/10.1080/19443994.2014.924032
53. Lakherwal D (2014) Adsorption of heavy metals: a review. Int J Environ Res Dev 4(1):41–48
54. Leaković S, Mijatović I, Cerjan-Stefanović Š, Hodžić E (2000) Nitrogen removal from fertilizer wastewater by ion exchange. Water Res 34(1):185–190. https://doi.org/10.1016/S0043-1354(99)00122-0
55. Lebrero R, Bouchy L, Stuetz R, Muñoz R (2011).Odor assessment and management in wastewater treatment plants: a review. Critical Rev Environ Sci Technol 41(10):915–950. https://doi.org/10.1080/10643380903300000
56. Lee CS, Robinson J, Chong MF (2014).A review on application of flocculants in wastewater treatment. Process Safe Environ Protect 92(6):489–508. https://doi.org/10.1016/j.psep.2014.04.010
57. Libralato G, Ghirardini AV, Avezzù F (2012) To centralise or to decentralise: an overview of the most recent trends in wastewater treatment management. J Environ Manag 94(1):61–68. https://doi.org/10.1016/j.jenvman.2011.07.010
58. Lin SH, Lo CC (1996) Treatment of textile wastewater by foam flotation. Environ Technol 17(8):841–849. https://doi.org/10.1080/09593331708616452
59. Lin SH, Peng CF (1994) Treatment of textile wastewater by electrochemical method. Water Res 28(2):277–282. https://doi.org/10.1016/0043-1354(94)90264-X
60. Lin SH, Shyu CT, Sun MC (1998) Saline wastewater treatment by electrochemical method. Water Res 32(4):1059–1066. https://doi.org/10.1016/S0043-1354(97)00327-8
61. Liu YJ, Gu J, Liu Y (2018) Energy self-sufficient biological municipal wastewater reclamation: present status, challenges and solutions forward. Bioresource Technol 269:513–519. https://doi.org/10.1016/j.biortech.2018.08.104
62. Loganathan BG, Ahuja S, Subedi B (2020) Synthetic organic chemical pollutants in water: origin, distribution, and implications for human exposure and health. In: Contaminants in our water: identification and remediation methods. American Chemical Society, pp 13–39. https://doi.org/10.1021/bk-2020-1352.ch002
63. Mahvi AH (2009) Application of ultrasonic technology for water and wastewater treatment. Iranian J Public Health 1–17.
64. Marcucci M, Ciabatti I, Matteucci A, Vernaglione G (2003) Membrane technologies applied to textile wastewater treatment. Ann N Y Acad Sci 984(1):53–64. https://doi.org/10.1111/j.1749-6632.2003.tb05992.x
65. Marder L, Bernardes AM, Ferreira JZ (2004) Cadmium electroplating wastewater treatment using a laboratory-scale electrodialysis system. Separation Purif Technol 37(3):247–255. https://doi.org/10.1016/j.seppur.2003.10.011
66. Martínez R, Vela N, Aatik AE, Murray E, Roche P, Navarro JM (2020) On the use of an IoT integrated system for water quality monitoring and management in wastewater treatment plants. Water 12(4):1096. https://doi.org/10.3390/w12041096
67. McNamara CJ, Anastasiou CC, O'Flaherty V, Mitchell R (2008) Bioremediation of olive mill wastewater. Int Biodeterioration Biodegradation 61(2):127–134. https://doi.org/10.1016/j.ibiod.2007.11.003

68. Mo J, Yang Q, Zhang N, Zhang W, Zheng Y, Zhang Z (2018) A review on agro-industrial waste (AIW) derived adsorbents for water and wastewater treatment. J Environ Manag 227:395–405. https://doi.org/10.1016/j.jenvman.2018.08.069
69. Mohammadi AR, Mehrdadi N, Bidhendi GN, Torabian A (2011). Excess sludge reduction using ultrasonic waves in biological wastewater treatment. Desalination 275(1–3):67–73. https://doi.org/10.1016/j.desal.2011.02.030
70. Mohammadi T, Esmaeelifar A (2004) Wastewater treatment using ultrafiltration at a vegetable oil factory. Desalination 166:329–337. https://doi.org/10.1016/j.desal.2004.06.087
71. Morseletto P (2020) Targets for a circular economy. Resources, Conservation Recycling, 153:104553. https://doi.org/10.1016/j.resconrec.2019.104553
72. Mulchandani A, Westerhoff P (2016) Recovery opportunities for metals and energy from sewage sludges. Bioresource Technol 215:215–226. https://doi.org/10.1016/j.biortech.2016.03.075
73. Murray A, Skene K, Haynes K (2017) The circular economy: an interdisciplinary exploration of the concept and application in a global context. J Bus Ethics 140(3):369–380. https://doi.org/10.1007/s10551-015-2693-2
74. Naumczyk J, Bogacki J, Marcinowski P, Kowalik P (2014) Cosmetic wastewater treatment by coagulation and advanced oxidation processes. Environ Technol 35(5):541–548. https://doi.org/10.1080/09593330.2013.808245
75. Obotey Ezugbe E, Rathilal S (2020) Membrane technologies in wastewater treatment: a review. Membranes 10(5):89 https://doi.org/10.3390/membranes10050089
76. Oztekin E, Altin S (2016) Wastewater treatment by electrodialysis system and fouling problems. Turkish Online J Sci Technol 6(1)
77. Paquay E, Clarinval AM, Delvaux A, Degrez M, Hurwitz HD (2000) Applications of electrodialysis for acid pickling wastewater treatment. Chem Eng J 79(3):197–201. https://doi.org/10.1016/S1385-8947(00)00208-4
78. Pavithra KG, Kumar PS, Christopher FC, Saravanan A (2017) Removal of toxic Cr (VI) ions from tannery industrial wastewater using a newly designed three-phase three-dimensional electrode reactor. J Phys Chem Solids 110:379–385. https://doi.org/10.1016/j.jpcs.2017.07.002
79. Pollard SJT, Fowler GD, Sollars CJ, Perry R (1992) Low-cost adsorbents for waste and wastewater treatment: a review. Sci Total Environ 116(1–2):31–52. https://doi.org/10.1016/0048-9697(92)90363-W
80. Prabu D, Parthiban R, Senthil Kumar P, Kumari N, Saikia P (2016) Adsorption of copper ions onto nano-scale zero-valent iron impregnated cashew nut shell. Desalination Water Treatment 57(14):6487–6502. https://doi.org/10.1080/19443994.2015.1007488
81. Prasannamedha G, Kumar PS, Mehala R, Sharumitha TJ, Surendhar D (2021) Enhanced adsorptive removal of sulfamethoxazole from water using biochar derived from hydrothermal carbonization of sugarcane bagasse. J Hazardous Mater 407:124825. https://doi.org/10.1016/j.jhazmat.2020.124825
82. Prieto-Sandoval V, Jaca C, Ormazabal M (2018) Towards a consensus on the circular economy. J Clean Prod 179:605–615. https://doi.org/10.1016/j.jclepro.2017.12.224
83. Rajasulochana P, Preethy V (2016) Comparison on efficiency of various techniques in treatment of waste and sewage water–a comprehensive review. Resource-Efficient Technol 2(4):175–184. https://doi.org/10.1016/j.reffit.2016.09.004
84. Rajkumar D, Kim JG (2006) Oxidation of various reactive dyes with in situ electro-generated active chlorine for textile dyeing industry wastewater treatment. J Hazardous Mater 136(2):203–212. https://doi.org/10.1016/j.jhazmat.2005.11.096
85. Rajkumar D, Palanivelu K (2004) Electrochemical treatment of industrial wastewater. J Hazardous Mater 113(1–3):123–129. https://doi.org/10.1016/j.jhazmat.2004.05.039
86. Rashid R, Shafiq I, Akhter P, Iqbal MJ, Hussain M (2021). A state-of-the-art review on wastewater treatment techniques: the effectiveness of adsorption method. Environ Sci Pollut Res 1–17. https://doi.org/10.1007/s11356-021-12395-x

87. Rathi BS, Kumar PS, Ponprasath R, Rohan K, Jahnavi N (2021) An effective separation of toxic arsenic from aquatic environment using electrochemical ion exchange process. J Hazardous Mater 412:125240. https://doi.org/10.1016/j.jhazmat.2021.125240
88. Renge VC, Khedkar SV, Pande SV (2012) Removal of heavy metals from wastewater using low cost adsorbents: a review. Sci Revs Chem Commun 2(4):580–584
89. Rice RG (1996) Applications of ozone for industrial wastewater treatment—a review. Ozone: Sci Eng 18(6):477–515. https://doi.org/10.1080/01919512.1997.10382859
90. Rubio J, Souza ML, Smith RW (2002) Overview of flotation as a wastewater treatment technique. Minerals Eng 15(3):139–155. https://doi.org/10.1016/S0892-6875(01)00216-3
91. Ruppert G, Bauer R, Heisler G (1993) The photo-Fenton reaction—an effective photochemical wastewater treatment process. J Photochem Photobiol Chem 73(1):75–78. https://doi.org/10.1016/1010-6030(93)80035-8
92. Salahi A, Mohammadi T, Rahmat Pour A, Rekabdar F (2009) Oily wastewater treatment using ultrafiltration. Desalination Water Treatment 6(1–3):289–298. https://doi.org/10.5004/dwt.2009.480
93. Saravanan A, Kumar PS, Varjani S, Karishma S, Jeevanantham S, Yaashikaa PR (2021) Effective removal of Cr (VI) ions from synthetic solution using mixed biomasses: Kinetic, equilibrium and thermodynamic study. J Water Process Eng 40:101905. https://doi.org/10.1016/j.jhazmat.2020.124825
94. Sarkar B, Chakrabarti PP, Vijaykumar A, Kale V (2006) Wastewater treatment in dairy industries—possibility of reuse. Desalination 195(1–3):141–152. https://doi.org/10.1016/j.desal.2005.11.015
95. Schweitzer L, Noblet J (2018) Water contamination and pollution. In: Green chemistry. Elsevier, pp 261–290. https://doi.org/10.1016/B978-0-12-809270-5.00011-X
96. Senthil Kumar P, Senthamarai C, Sai Deepthi ASL, Bharani R (2013) Adsorption isotherms, kinetics and mechanism of Pb (II) ions removal from aqueous solution using chemically modified agricultural waste. Canadian J Chem Eng 91(12):1950–1956. https://doi.org/10.1002/cjce.21784
97. Sgroi M, Vagliasindi FG, Roccaro P (2018) Feasibility, sustainability and circular economy concepts in water reuse. Current Opinion Environ Sci Health 2:20–25. https://doi.org/10.1016/j.coesh.2018.01.004
98. Sharshir SW, Algazzar AM, Elmaadawy KA, Kandeal AW, Elkadeem MR, Arunkumar T et al (2020) New hydrogel materials for improving solar water evaporation, desalination and wastewater treatment: a review. Desalination 491:114564. https://doi.org/10.1016/j.desal.2020.114564
99. Shen KW, Li L, Wang JQ (2020) Circular economy model for recycling waste resources under government participation: a case study in industrial waste water circulation in China. Technol Econ Dev Economy 26(1):21–47. https://doi.org/10.3846/tede.2019.11249
100. Shete BS, Shinkar NP (2013) Dairy industry wastewater sources, characteristics & its effects on environment. Int J Current Eng Technol 3(5):1611–1615. https://doi.org/10.1021/ie501210j
101. Sivaranjanee R, Saravanan A (2018) Carbon sphere: synthesis, characterization and elimination of toxic Cr (VI) ions from aquatic system. J Indus Eng Chem 60:307–320. https://doi.org/10.1016/j.jiec.2017.11.017
102. Slater CS, Ahlert RC, Uchrin CG (1983) Applications of reverse osmosis to complex industrial wastewater treatment. Desalination 48(2):171–187. https://doi.org/10.1016/0011-9164(83)80015-0
103. Smith KM, Fowler GD, Pullket S, Graham ND (2009) Sewage sludge-based adsorbents: a review of their production, properties and use in water treatment applications. Water Res 43(10):2569–2594. https://doi.org/10.1016/j.watres.2009.02.038
104. Smol M, Adam C, Preisner M (2020) Circular economy model framework in the European water and wastewater sector. J Mater Cycles Waste Manag 1–16. https://doi.org/10.1007/s10163-019-00960-z

105. Šostar-Turk S, Petrinić I, Simonič M (2005) Laundry wastewater treatment using coagulation and membrane filtration. Resour Conservation Recycling 44(2):185–196. https://doi.org/10.1016/j.resconrec.2004.11.002
106. Thekkudan VN, Vaidyanathan VK, Ponnusamy SK, Charles C, Sundar S, Vishnu D, et al (2016) Review on nanoadsorbents: a solution for heavy metal removal from wastewater. IET Nanobiotechnol 11(3):213–224. https://doi.org/10.1049/iet-nbt.2015.0114
107. Topare NS, Attar SJ, Manfe MM (2011) Sewage/wastewater treatment technologies: a review. Sci Revs Chem Commun 1(1):18–24
108. Voulvoulis N (2018) Water reuse from a circular economy perspective and potential risks from an unregulated approach. Current Opinion Environ Sci Health 2:32–45. https://doi.org/10.1016/j.coesh.2018.01.005
109. Vourch M, Balannec B, Chaufer B, Dorange G (2008) Treatment of dairy industry wastewater by reverse osmosis for water reuse. Desalination 219(1–3):190–202. https://doi.org/10.1016/j.desal.2007.05.013
110. Wuhrmann K, Mechsner KL, Kappeler TH (1980) Investigation on rate—determining factors in the microbial reduction of azo dyes. Euro J Appl Microbiol Biotechnol 9(4):325–338. https://doi.org/10.1007/BF00508109
111. Xiang W, Zhang X, Chen J, Zou W, He F, Hu X, et al (2020) Biochar technology in wastewater treatment: a critical review. Chemosphere 252:126539. https://doi.org/10.1016/j.chemosphere.2020.126539
112. Xu P, Zeng GM, Huang DL, Feng CL, Hu S, Zhao MH et al (2012) Use of iron oxide nanomaterials in wastewater treatment: a review. Sci Total Environ 424:1–10. https://doi.org/10.1016/j.scitotenv.2012.02.023
113. Zaharia C, Suteu D, Muresan A, Muresan R, Popescu A (2009) Textile wastewater treatment by homogenous oxidation with hydrogen peroxide. Environ Eng Manag J 8(6):1359–1369
114. Zhao L, Dai T, Qiao Z, Sun P, Hao J, Yang Y (2020) Application of artificial intelligence to wastewater treatment: a bibliometric analysis and systematic review of technology, economy, management, and wastewater reuse. Process Safe Environ Protect 133:169–182. https://doi.org/10.1016/j.psep.2019.11.014

Printed in the United States
by Baker & Taylor Publisher Services